Low Transformation Temperature Welding Materials

Properties, Weldability, and Applications

Ninshu Ma, Kazuo Hiraoka, Zhongyuan Feng, and Hiroshi Yajima

CRC Press
Taylor & Francis Group
Boca Raton London New York

CRC Press is an imprint of the
Taylor & Francis Group, an **informa** business

Designed cover image: Ninshu Ma, Kazuo Hiraoka, Zhongyuan Feng, and Hiroshi Yajima

First edition published 2026
by CRC Press
2385 NW Executive Center Drive, Suite 320, Boca Raton FL 33431

and by CRC Press
4 Park Square, Milton Park, Abingdon, Oxon, OX14 4RN

CRC Press is an imprint of Taylor & Francis Group, LLC

ISBN: 9781032944838 (hbk)
ISBN: 9781032990163 (pbk)
ISBN: 9781003601845 (ebk)

DOI: 10.1201/9781003601845

Typeset in Palatino
by codeMantra

Access the Support Material: www.routledge.com/9781032944838

Contents

About the Authors

Ninshu Ma was born in 1961 and received his PhD in 1994 from The University of Osaka, Japan. He is a professor in the Joining and Welding Research Institute and the director of the East-Asian Center of Academic Initiatives at The University of Osaka, Japan. He is an associate editor of *Ships and Offshore Structure*, editor of *China Welding*, and associate editor-in-chief of *Advanced Materials Joining*. He was the editor-in-chief of *The Journal of the Japan Society of Technology of Plasticity*. He has published more than 300 journal articles and 6 books/chapters. He has chaired/co-chaired several international conferences. His research interests include computational welding mechanics and metal forming mechanics, development of computing/measuring methods, and material constitutive models for various fusion/solid-state joining methods, as well as additive manufacturing.

Kazuo Hiraoka was born in 1948 and received his Dr.-Eng. in 1997 from The University of Osaka, Japan. He was a guest (2016–2024) and visiting (2009–2016) professor in the Joining and Welding Research Institute at The University of Osaka, Japan. Until 2009, he was the leader of the Welding Metallurgy Group at the National Institute for Materials Science (NIMS). He has published more than 100 journal articles and 3 books/chapters. His research interests include diagnostics of welding arc, development of arc welding process, and welding technology for 780–980 MPa class high-strength steels with LTT materials.

Zhongyuan Feng (born in 1988) received his PhD in 2021 from The University of Osaka, Japan. He has published more than 15 journal articles. His research interests include weldability of metallic materials, residual stress and fatigue performance of welded joints, and friction stir welding technology for high-melting-point materials.

Hiroshi Yajima was born in Tokyo in 1936 and received his Dr. Eng. in 1981 from The University of Tokyo, Japan. He worked at the Technical Headquarters of Mitsubishi Heavy Industries Ltd. from 1960 through 1994. He was a professor at the Graduate School of Engineering, Hiroshima University from 1994 through 2000, and he was a professor at the Graduate School of Engineering, Nagasaki Institute of Applied Science from 2000 through 2015. At present, he is an emeritus professor at the Nagasaki Institute of Applied Science. He has published about 500 journal articles. His research interests include fatigue strength and unstable fracture strength of large welded structures.

Preface

Tensile residual stress in welded joints is one of the main causes of weld cracking, reducing fatigue strength and its life, stress corrosion cracking, brittle fracture, and induced damage. Conversely, compressive residual stress has a positive effect on the above properties of welded joints.

Low transformation temperature (LTT) welding materials utilize the thermal expansion at relatively lower temperatures induced by martensitic transformation to generate compressive residual stress in welded joints, which can improve fatigue and corrosion performance as well as brittle fracture resistance.

The LTT welding materials were first investigated in the 1980s, and attention was paid by many researchers in the welding field. From 2010, LTT welding materials with various chemical compositions were designed with or without consideration of welding dilutions, and their properties were investigated through experiments. Recently, special attention has been paid to the practical application for the successful prevention of fatigue cracks occurring at the toe of fillet welds, especially boxing filled welds in ship structures. Another potential application is surface cladding and wire-arc additive manufacturing for the improvement of stress corrosion performance of pipe and plate structures. The good characteristics of compressive residual stress, high strength, and good toughness at low temperatures may extend their application fields further. On the other hand, the ultra-high-strength steels and hot stamping steels widely used in automotive bodies have similar characteristics to LTT welding materials.

How to actively use LTT materials with martensitic transformation and for compressive residual stress generation is of more interest. In this book, the authors would like to provide readers with the following information.

- A review of LTT welding materials; design of chemical compositions for LTT welding materials; properties and weldability of LTT welding materials; compressive residual stress; stress concentration; practical application examples of LTT welding materials for the prevention of fatigue cracking.
- Using experimental and numerically simulated data, photos, graphs, and tables to show the information in this book for easy understanding and visual quantitative valuation.
- Understanding the mechanism of LTT welding materials for compressive residual stress generation; referring to the demonstrated welding procedures; extending the applications based on the knowledge in this book.

In Chapter 1, the definition of LTT welding materials is given. As examples, various LTT welding materials such as Fe–Cr–Ni-based alloy, Fe–Ni-based alloy, Fe–Cr-based alloy, Fe–Cr–Mn based alloy, Fe–Mn-based alloy, and Fe–Mn–Ni-based alloy, designed by researchers around the world, are reviewed. Then general characteristics of LTT welding materials are introduced. Finally, key factors in LTT welding materials design for applications are discussed.

In Chapter 2, several new LTT materials with different chemical compositions are introduced. Their mechanical, physical, and corrosion properties and microstructures of LTT weld metals with different dilutions corresponding to different welding conditions in various welded joints are measured. The design concept for LTT welding materials with consideration of different dilutions is proposed. Based on this concept, the 16%Cr–8%Ni LTT welding material was developed. Its weld metals with different dilution rates during solidification showed the preferred FA mode for welding crack prevention, which was identified by in-situ observation using XRD methods. The martensite start (Ms) temperature and phase transformation expansion strain of the LTT weld metals satisfy the conditions to generate large compressive residual stress. The LTT weld metals had high yield strength, high tensile strength, good elongation, acceptable fracture toughness, and good corrosion resistance for practical applications. The flux-cored 16%Cr–8%Ni LTT welding wire has excellent weldability.

In Chapter 3, the different LTT welding materials induced residual stress and its generation mechanism in various welded joints, including a simple constraint bar, a bead-on-plate weld, butt welds, fillet welds, and boxing fillet welds, are discussed based on numerical analysis and experimental measurements. Then, the influence of plate size and welding sequence on the welding residual stress is also introduced.

In Chapter 4, the application of LTT welding materials to fatigue crack prevention of boxing fillet-welded joints with the elongated bead is focused. Especially, repair and reinforcement welding procedures in all welding positions are described in detail. Both finite element analysis and experimental measurement showed that a large compressive residual stress in the elongated bead welded in all positions is generated. The fatigue test results with large-size specimens fabricated at a shipyard are shown for applicability and reliability. Furthermore, considering the plate thickness combinations of welded joints in ship structures, various numerical experiments for the analysis of stress concentration and residual stress on the LTT elongated bead, as well as stress intensity factors at the root of boxing fillet-welded Joints, were conducted for the practical application of the LTT welding material.

In Chapter 5, the application possibilities of LTT welding materials in new fields such as wire-arc additive manufacturing and surface cladding are demonstrated.

In the appendix, digital data of the finite element mesh of a butt weld, a fillet weld, and a boxing fillet weld with an elongated bead are supplied to readers for their free use.

Keywords: Low Transformation Temperature Welding Materials; Weldability; Properties; Compressive Residual Stress; Stress Intensity Factor; Fatigue Life; Fatigue Crack Prevention; Corrosion Resistance; Repair Welding; Reinforcement Welding; All Welding Positions; Welding Process; Wire-Arc Additive Manufacturing; Surface Cladding

Ninshu Ma
Kazuo Hiraoka
Zhongyuan Feng
Hiroshi Yajima

List of Abbreviations and Symbols

Chapter 1

EL	elongation
HT780	high strength steel with the minimum tensile strength of 780 MPa
HT570	high strength steel with the minimum tensile strength of 570 MPa
LTT	low transformation temperature
Mf	martensite finish temperature
Ms	martensite start temperature
Nf	fatigue failure cycles
R	ratio of the minimum stress to the maximum stress in fatigue test
S355J2	low-carbon steel with the minimum yield strength of 355 MPa and the tensile strength of 470–680 MPa
S960QL	high-strength steel with the minimum tensile strength of 960 MPa
S–N	stress-cycle curve for fatigue failure
σ_x	(residual) stress in x-direction
σ_y	(residual) stress in y-direction
Tekken test	y-groove weld cracking test
TS	tensile strength
YS	yield strength

Chapter 2

$\bar{V}$	average volume of the newly formed martensite due to temperature variation
$\Delta G_y^{\delta-\alpha'}$	driving force (free energy) for the formation of martensite
$2a$	central through-notch length in the low-temperature fracture toughness test
$2B$	specimen width in the low-temperature fracture toughness test
A, AF mode	solidification mode with the primary phase austenite
a_γ	linear expansion coefficient of austenite

α_m	linear expansion coefficient of martensite
Class-NK	ship classification society of Japan (Nippan Kaiji Association)
Cr_{eq}	Cr equivalent
CTOD	crack tip opening displacement
$\Delta\varepsilon$	apparent martensitic transformation expansion strain
$\varepsilon_{\gamma-m}$	strain due to the lattice volume difference between austenite and martensite
EL	elongation
ESW	electroslag welding
FA mode	solidification mode with the primary phase ferrite
FC value	parameter representing crack propagation
FCW	flux-cored wires
FISCO Test	weld cracking test for solidification crack, C-restraint weld cracking test
f_m	volume fraction of martensite
formastor®	transformation measuring apparatus
FSM	field signature method
f_γ	volume fraction of austenite
γ	austenite
GMAW	Gas metal arc welding
Gumbel probability	probability distribution of the maximum (or the minimum) of a number of samples of various distributions
HV	Vickers hardness
j	proportionality constant
JIS	Japanese Industrial Standards
KA36	hull structural steel of Nippan Kaiji Association (Class-NK)
K_C	fracture toughness
KD36	hull structural steel of Nippan Kaiji Association (Class-NK)
MAG	metal active gas arc welding (GMAW)
mdd	corrosion-induced weight reduction to mg per 100 cm^2 and per day (mg/100 cm^2/day)
Mf	martensite finish temperature
MIG	metal inert gas arc welding (GMAW)
Ms	martensite start temperature
Ms_{calc}	approximated martensitic transformation start temperature calculated from measured thermal expansion history curve during cooling
Ms_{est}	estimated martensitic transformation start temperature using chemical compositions

MT	magnetic particle testing
Ni_{eq}	Ni equivalent
P_{CM}	cracking parameter of material
ppm	parts per million
ppt	pair per thousand
RH	relative humidity
SAW	submerged arc welding
Schaeffler diagram	a graphical to describe the stable phases of steels at room temperature, based on their chemical compositions.
SEM	scanning electron microscope
σ_g	gross stress
SM490	steel with the minimum tensile strength of 490 MPa
SMAW	shielded metal arc welding
σ_n	net fracture stress
Tekken test	y-groove weld cracking test for cold cracking
TIG	tungsten inert gas arc welding (GTAW)
TS	tensile strength
$vE_{0°C}$	absorbed energy at 0°C in V-notch Charpy impact test
vT_{rE}	50% absorbed energy transition temperature
XRD	X-ray diffraction
YS	yield strength
ΔD	apparent martensitic transformation expansion displacement
ΔH	corrosion depth
ΔH_m	maximum corrosion depth
ΔH_{me}	predicted maximum corrosion depth
$\Delta\sigma$	stress range
$\Delta\sigma_{gross}$	gross stress range

Chapter 3

$\overline{\sigma}$	equivalent stress
[B]	corresponding matrix between strain and displacement
[C]	corresponding matrix between strain and stress
[ΔC]	increment of matrix [C]
$\{\Delta\sigma\}$	stress increment vector
$\{du\}$	displacement increment vector
$\{d\varepsilon\}$	strain increment vector
$\{d\varepsilon^e\}$	elastic strain increment vector

$\{d\varepsilon^{p}\}$	plastic strain increment vector
$\{d\varepsilon^{ph}\}$	phase transformation-induced strain increment vector
$\{d\varepsilon^{T}\}$	thermal strain increment vector
$\{\varepsilon\}$	strain vector
$\{\sigma'\}$	deviation stress
$\{\sigma\}$	stress vector
α phase	state defined as a body-centered tetragonal structure
Ac1	start temperature of phase transformation from ferrite to austenite during heating
Ac3	finish temperature of phase transformation from ferrite to austenite during heating
BM	base metal
CCT diagram	continuous cooling transformation diagram
CTE	coefficient of thermal expansion
$\{\Delta\varepsilon^{tp}\}$	phase transformation-induced non-volumetric plastic strain increment
$\{\Delta\varepsilon^{vol}\}$	phase transformation-induced volume expansion strain increment
dt	time increment
E	Young's modulus
FEA	finite element analysis
FEM	finite element method
γ phase	state defined as a *face-centered cubic* structure
HAZ	heat-affected zone
JMatPro	thermodynamic software
K	coefficients of transformation plasticity
KA36	hull structural steel of Nippan Kaiji Association (Class-NK)
Mf	martensite finish temperature
Ms	martensite start temperature
σ_Y	yield stress
T	temperature
T	time or previous time
$t + dt$	current time which is the previous time t plusing time increment dt
T_{ref}	reference temperature
υ	Poisson's ratio
WM	weld metal
α	transient thermal expansion coefficient
α_0	static thermal expansion coefficient
$\Delta\varepsilon_{eq}{}^{T}$	equivalent strain increment
$\Delta\varepsilon_{th}$	increment of thermal strain
$\Delta\lambda$	coefficient between plastic strain vector and stress change in yield surface

Chapter 4

a	parameter of penetration depth at the bottom of the gusset plate end
A-conv. Joint	boxing fillet-welded joint using conventional welding wire
AD-joint.	joints treated by toe grinding in A-Conv. joints
b	parameter of penetration depth at the bottom of the gusset plate sides
B-LTT joint	boxing fillet-welded joint using 10%Cr-10%Ni LTT wire
c	specific heat
$\Delta\sigma$	stress range, the difference between stress maximum and stress minimum in a stress cycle
$\Delta\sigma_{2e6}$	stress range to failure in $N_f = 2 \times 10^6$
ε^e	elastic strain
EL	elongation
ε^p	plastic strain
ER-10Cr10Ni joint	boxing fillet-welded joint overlayed elongated bead weld by using LTT wire in A-Conv. Joint
ER-16Cr8Ni joint	boxing fillet-welded joint overlayed A-Conv. joint with elongated bead weld by using 16%Cr–8%Ni LTT wire
ε^{th}	thermal strain
ε^{tr}	transformation strain
F	Flat welding position
FAT	fatigue class
FCW	flux-cored wire
FE analyses	finite element analyses
FR-10Cr10Ni joint	boxing fillet-welded joint overlaying elongated bead weld by using 10%Cr–10%Ni LTT wire in A-Conv. joint with the machined groove of the angle of 60° and the depth of 5 mm at the bottom of the gusset plate end
GR-16Cr8Ni joint	boxing fillet-welded joint overlaying elongated bead weld by using 16%Cr–8%Ni LTT wire in A-Conv. joint with the air-arc gauged groove of the depth of 5 mm at the bottom of the gusset plate end
H	horizontal welding position
H_{bead}	leg length of elongated bead
HiFIT	high-frequency impact treatment
HV2	Vickers hardness under test load 2 kgf
IIW	International Institute of Welding
JWRIAN	FEM program developed in Osaka University
K	stress intensity factor
KA36	hull structural steel of Nippan Kaiji Association (Class-NK)
λ	thermal conductivity

m	slope of the fatigue strength curves for details assessed on the basis of normal stresses
Mf	martensite finish temperature
Ms	martensite start temperature
N_f	number of cycles to failure in fatigue test
$N_{f150\,\mathrm{MPa}}$	number of cycles to failure in stress range 150 MPa
$N_{f\mathrm{Conv.mean}}$	mean fatigue life of the conventional boxing fillet-welded joint
OH	overhead welding position
R	stress ratio
RF specimen	reinforced specimen
RP specimen	repaired specimen
σ_F	stress concentration-induced stress
σ_R	residual stress
σ_T	evaluation stress
STDV	standard deviation
t_p	main plate thickness
TRIP	transformation-induced plasticity
t_s	stiffener plate thickness
TS	tensile strength
UIT	ultrasonic impact treatment
${}_VE_{0°C}$	absorbed energy at test temperature = 0°C in V-notched Charpy impact test
VU/L	vertical upward welding position at lower side of stiffener
VU/U	vertical upward welding position at upper side of stiffener
W_{bead}	elongated bead width
XRD	X-ray diffractometer
YS	yield strength

Chapter 5

10Cr10Ni	10%Cr and 10%Ni
16Cr8Ni	16%Cr and 8%Ni
HAZ	heat-affected zone
KA36	hull structural steel of Nippan Kaiji Association (Class-NK)
Ms	martensite start temperature
SUS304	stainless steel 304
SUS308L	stainless steel 308L
WAAM	wire-arc additive manufacturing
XRD	X-ray diffraction
σ_C	stress component in the circumferential direction
σ_L	stress component in the longitudinal direction

1

Review of Various LTT Welding Materials

1.1 Definition and Purpose of Low Transformation Temperature Welding Materials

While welding offers efficiency and cost-effectiveness, thermal contraction and geometric constraints induce residual stress, particularly tensile stress, in and around the weld zone. This tensile stress can shorten the fatigue strength of welded joint, especially when using high-strength base metals. Thus, developing welding alloys that can mitigate tensile residual stress is highly desirable across various industries.

In the past decades, researchers have been dedicated to finding such "smart" alloys and have finally developed low transformation temperature (LTT) welding materials. LTT welding materials are a type of welding consumable that undergo martensitic transformation during cooling (with a martensite start temperature generally less than 400°C) and result in volume expansion that can counteract or even reverse the contraction that occurs during the welding process, leading to reduction of tensile residual stress or even production of compressive residual stress in the weld area.

Figure 1.1.1 is a commonly featured graph in numerous publications, illustrating a comparison between an LTT weld metal and a conventional weld metal [1,2]. As depicted by the dashed line in Figure 1.1.1a, the martensite start (Ms) temperature for conventional welding materials is approximately 550°C. Following the completion of the transformation, as the temperature decreases, contraction strain ensues. Should the material's contraction from its service temperature—namely, room temperature—be impeded by surrounding constraints, tensile residual stress will be generated, as illustrated in Figure 1.1.1b. Conversely, when the Ms temperature is around 200°C and the transformation finishes near room temperature, the welding metal remains in an expanded state upon the completion of welding. This condition allows for the introduction of compressive residual stress.

The purposes of LTT welding materials are listed in the following:

1. **Reduce Welding Residual Stress:** The martensitic transformation expansion during the welding process with LTT welding materials can regulate the residual stress of the welded joint, reducing the tensile residual stress at the weld toe and even producing compressive residual stress, as shown in Figure 1.1.2, where wires A, B, and

DOI: 10.1201/9781003601845-1

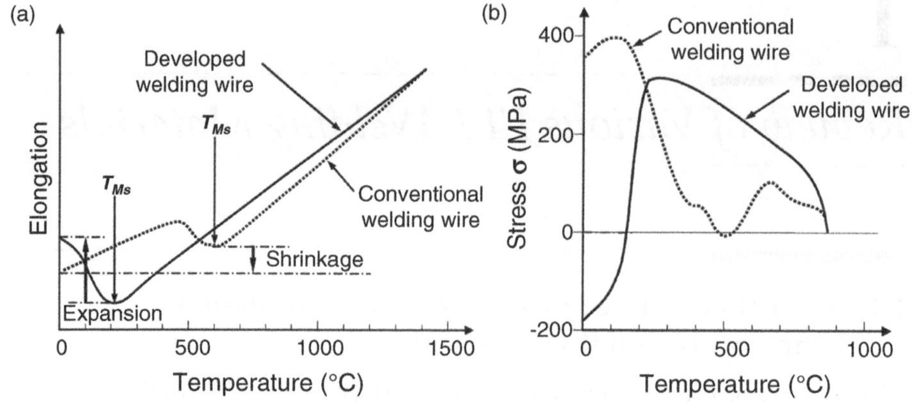

FIGURE 1.1.1
Characteristics of LTT materials: (a) transformation of weld metal during unconstrained cooling and (b) development of stress in the constraint bar model [1,2].

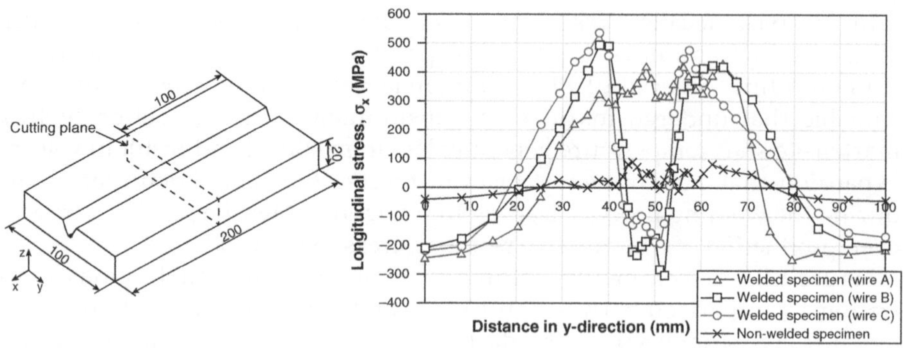

FIGURE 1.1.2
Comparison of longitudinal residual stress measured by the contour method of welded and non-welded specimens [3].

C are the conventional (Kobelco: MG-S50), 10%Cr–10%Ni LTT, and 20%Cr–8.6%Ni LTT, respectively [3]. In the case of wire A, the tensile residual stress at the welding surface is formed, and the peak tensile stress is higher than 400 MPa. Further away from the area, compressive stress is induced to maintain torque balance. On the other hand, both wires B and C induce compressive stress, with values even lower than –200 MPa in the weld.

2. **Extend Fatigue Life of Welded Joints:** LTT welding materials improve the fatigue performance of welded joints through metallurgical means, becoming a novel method to enhance the fatigue performance of welded joints [4]. Figure 1.1.3 shows the comparison of fatigue life among different welded joints. The LTT welded joint was

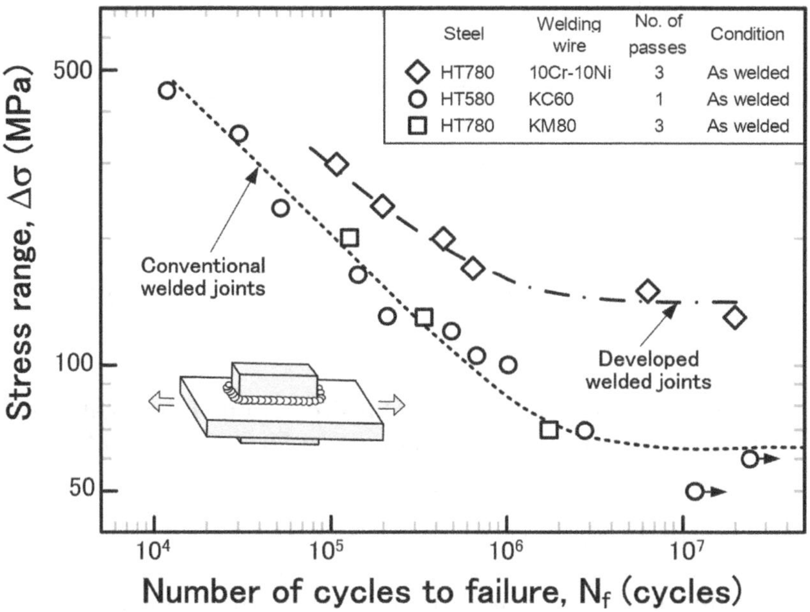

FIGURE 1.1.3
S–N curves of boxing fillet-welded joints [4].

fabricated using HT780 steel plate and 10%Cr–10%Ni welding wire. Conventional welded joints of HT580 and HT780 steel were produced with autogenous welding materials (JKW: KC60 and KM80), respectively. The fatigue properties of the conventional HT580 and HT780 welded joints were nearly identical, with a fatigue limit of 65 MPa observed for the HT580 conventional joint. The fatigue limit of the LTT welded joint is about twice that of the conventional joints.

3. **Welding Distortion Control:** The transformation plasticity associated with the formation of martensite, influenced by the stresses that develop as a weld cools, has been proven to effectively reduce distortion in the final weld [5]. Figure 1.1.4 shows the schematics of welding distortion in the conventional and LTT welded joints. Akyel et al. [6] utilized the LTT effect to control the welding distortion for high-alloy steels in electron beam welding. Their study found that the conventional weld specimens contracted after welding, resulting in an average distortion angle of 0.55°. In contrast, the LTT weld specimens showed a bend in the negative direction, with an average distortion angle of –0.08°. With the proportion of LTT microstructure in the upper area of the weld seam, sufficient compressive stress could be introduced into the specimen through volume expansion, so that compensation of the distortion could take place, and even overcompensation was achieved.

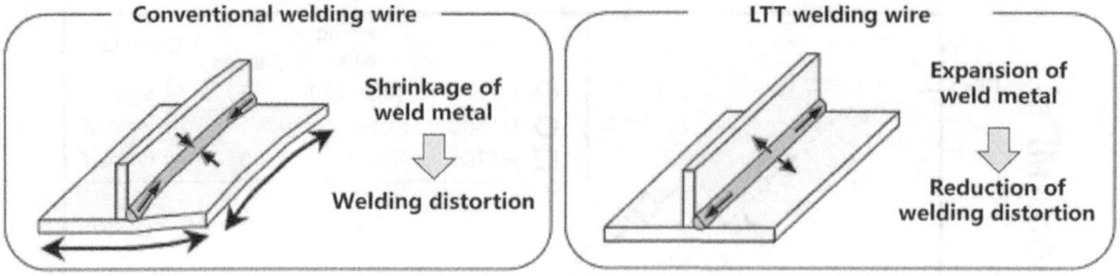

FIGURE 1.1.4
Distortion of mild steel plates joined with a single-pass weld using two types of welding consumables [7,8].

4. **Energy-Saving and Environmental Protection:** To mitigate tensile residual stress and enhance the fatigue performance of welded joints, a variety of post-weld treatments have been put forward. These include stress relief heat treatment, ultrasonic peening, modification of structural configurations, and thermal tensioning techniques. However, the implementation of these techniques is often hampered by technical complexity and/or prohibitive costs. Compared to other post-weld treatment methods, using LTT welding materials is not only efficient and cost-effective but also energy-saving and environmentally friendly.
5. **Suitable for High Strength Steel Welding:** LTT welding materials are particularly suitable for the welding of high strength steels because high strength steels are more sensitive to residual stress, and their fatigue strength is mainly related to the stress distribution at the weld toe [9].

In summary, LTT welding materials are defined as welding materials that undergo a martensitic transformation close to room temperature during cooling. Their main purpose is to reduce welding residual stress and enhance the fatigue strength of welded joints, and they are particularly suitable for the welding of high strength steel structures.

1.2 Mechanism of Stress Reduction: Why Martensite?

During welding, phase transformations are significantly influenced by the chemical composition and the cooling rate in regions undergoing (partial) austenitization. These variables directly impact transformation temperatures, leading to variations in microstructures and mechanical properties. This complex interplay—as detailed by Nitschke-Pagel and Wohlfahrt [10], Jones and Alberry [11], and Satoh [12]—ultimately affects residual stress formation.

1.2.1 Martensitic Transformation Temperatures

Early studies by Bühler and Scheil [13] in 1933 on iron-based alloys with varying nickel content explored the impact of Ms temperature on residual stress after heat treatment. They attributed this stress to a competition between hindered shrinkage and volume expansion during phase transformations, with quenching playing a crucial role. Their findings indicated that Ms temperature is a primary determinant of residual stress levels, while the position of martensite finish (Mf) temperature relative to ambient temperature is another influential factor.

Murata et al. [14] and Wang et al. [15] investigated the impact of phase transformation temperature on residual stress distribution in weld zones. Murata et al.'s results (Figure 1.2.1) demonstrated that LTT weld metals with proper Ms temperature can not only reduce tensile residual stress to zero but also induce compressive residual stress upon cooling to room temperature. This peak compressive stress occurs around 200°C, where the volume

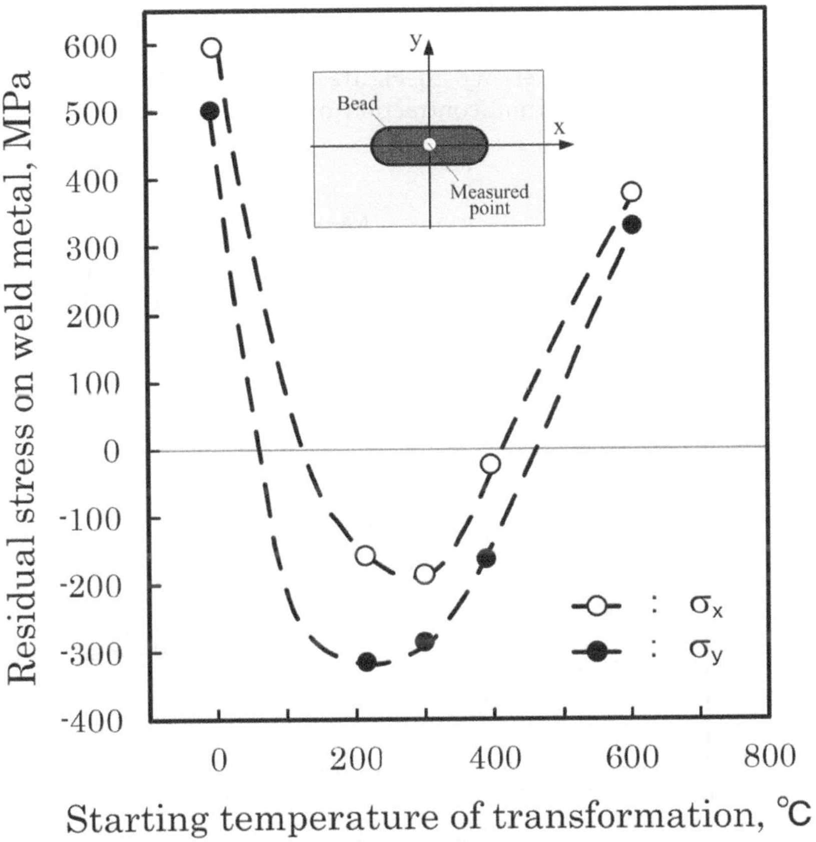

FIGURE 1.2.1
Relation between residual stress (σ_x, σ_y) of weld metal and starting temperature of transformation [14,16].

expansion accompanying the martensitic transformation is maximized. Below this temperature, incomplete transformation and volume expansion limit the reduction of tensile residual stress. In the absence of phase transformation, as with austenitic filler alloys, tensile stress increases progressively due to uninterrupted thermal contraction.

1.2.2 Effect of Phase Transformation on Residual Stress

The transformation of austenite upon cooling can occur through either a reconstructive or displacive mechanism. In the reconstructive transformation, atomic bonds break and reform, leading to the formation of allotriomorphic ferrite and/or pearlite. In contrast, displacive transformation involves a homogeneous deformation of the crystal lattice without bond breakage. Rapid cooling rates or heavy alloying can favor displacive transformation, resulting in the formation of acicular ferrite, bainite, or martensite. These phase transformations are accompanied by significant shear and volume expansion.

Jones and Alberry [11] investigated the mechanisms of phase transformations on welding-induced residual stress, confirming the principles outlined by Bühler and Scheil. As shown in Figure 1.2.2, the fully austenitic steel displays a near-linear thermal contraction profile. In contrast, bainitic and

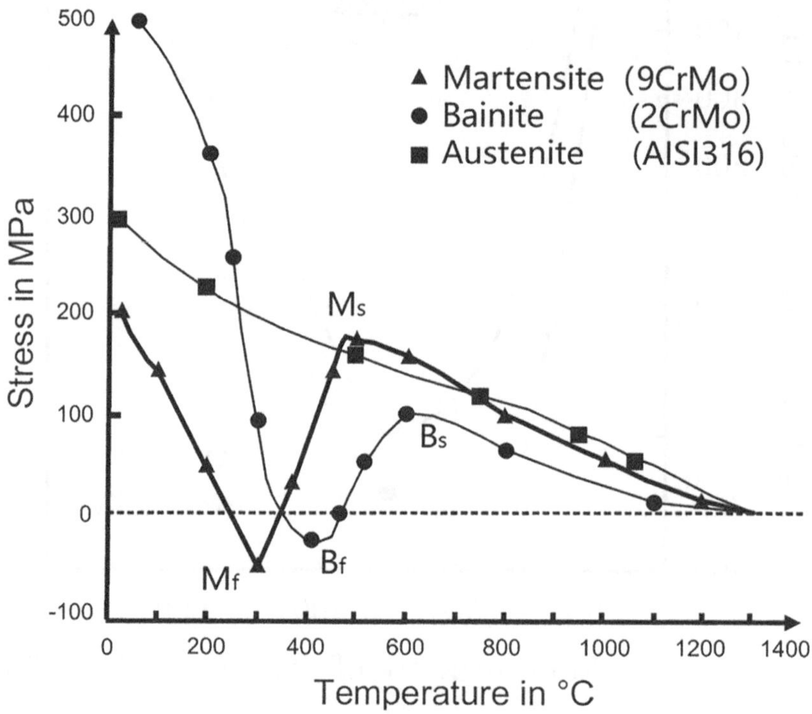

FIGURE 1.2.2
Stress history during the cooling of austenite, bainite, and martensite [11,17].

martensitic steels exhibit a reduced contraction rate during transformation, attributed to the relaxation of residual stress. Nevertheless, subsequent cooling to ambient temperature negates this beneficial effect. These observations indicate that the final stress state of a welded component is not only determined by the transformation temperature but also influenced by the temperature at which the transformation is fully completed.

Bhadeshia HKDH [18] further explained this phenomenon: on the one hand, the difference in thermal expansion coefficients between austenite and ferrite leads to greater volume expansion at lower temperatures. On the other hand, the face-centered cubic (FCC) lattice of austenite, with its lower yield strength (YS), allows for better stress compensation compared to the body-centered cubic (BCC) lattice of ferrite, which exhibits lower shrinkage compensation and higher stress gradients.

To maximize the stress-reducing effect of phase transformations, the transformation should ideally finish close to ambient temperature, allowing for the greatest deformation and transformation plasticity. While martensite offers significant advantages, carbon-rich martensite in low-alloy steels is undesirable due to its poor mechanical properties. Therefore, the LTT concept focuses on achieving martensitic structures with LTTs through careful alloy design.

In summary, the LTT concept leverages martensite for the following reasons:

Flexibility: Martensitic transformation temperatures can be easily adjusted through alloying elements, avoiding high carbon content.

Reliability: Under practical conditions, martensitic transformation is independent of the cooling rate; instead, it is primarily governed by temperature.

Effectiveness: Martensite offers the greatest potential for stress reduction compared to ferrite and bainite.

1.3 Classification of LTT Welding Materials

Typically, the chemical compositions of LTT welding consumables include the following alloy elements: C, Cr, Ni, Mn, Si, and even Mo. Other additions are generally less than 0.05 wt%. Consequently, LTT alloys can be categorized as follows:

1.3.1 Fe–Cr–Ni-Based Alloy

Following groundbreaking research on residual stress relaxation through transformation in a constrained body [11], Murata et al. [19] developed an iron-based alloy using chromium and nickel as the primary alloying

TABLE 1.3.1

Chemical Composition (wt%) and Ms Temperature of Fe–Cr–Ni-based LTT Weld Metals

LTT Alloy	C	Si	Mn	Cr	Ni	Mo	Other	Ms (°C)	References
10Cr–10Ni	0.025	0.32	0.7	10	10	0.13		180	[19]
L2	0.04	0.17	0.27	11.69	10.01	0.04		80	[15]
L3	0.04	0.17	0.3	10.51	9.62	0.05		138	[15]
L4	0.07	0.23	1.25	9.1	8.46	0.05		191	[15]
L5	0.08	0.17	1.35	7.78	6.88	0.06		242	[15]
L6	0.07	0.23	1.3	6.3	5.29	0.04		287	[15]
L7	0.09	0.47	1.51	5.07	4.43	0.06		325	[15]
L8	0.08	0.45	1.5	3.91	3.23	0.04		408	[15]
LTTE	0.07	0.2	1.3	9.1	8.5			200	[20]
B206	0.01	0.4	1.8	12.5	6.7	2.5	0.5Cu	<150	[21]
13Cr/LC35	0.04	0.5	0.8	12.3	7.3	2.2		<150	[21]
D	0.047	0.4	1.5	11.1	8.6	0.3		94	[22]
B5	0.04	0.19	0.86	13	1.7	0.04		300	[23]
C5	0.05	0.22	0.41	3	13.2	0.35		270	[23]

elements (10%Cr–10%Ni), with an Ms temperature of approximately 180°C. Table 1.3.1 lists the chemical composition and Ms temperature of Fe–Cr–Ni-based LTT weld metals from the published studies.

Most of the LTT alloys designed over the past two decades have featured a combination of Cr and Ni, with the majority also having low carbon content [22,24,25]. According to published data, LTT weld metals typically contain Ni in the range of 3%–12% and Cr in the range of 3%–16%, while Ms temperature can vary between 500°C and –25°C. Figure 1.3.1 shows Ms temperature as a function of Cr and Ni content only, neglecting the effects of other alloying elements [17,26]. The temperature range plotted is from 450°C to 25°C. The figure indicates that Ms temperature decreases with an increase in the Ni and/or Cr content.

1.3.2 Fe–Ni-based Alloy

Nickel is a popular choice for alloying because it is known to enhance toughness, provided that the cost of nickel is not prohibitively high. Francis et al. [27] designed an LTT alloy with nickel as a major alloying element; however, its toughness was not significantly superior to that of alloys containing a combination of chromium and nickel. Kannengiesser et al. [28] developed a 12%Ni LTT alloy that exhibited a cellular structure of martensite with pronounced segregation areas forming a network-like pattern across the weld, and it displayed several cracks identified as solidification cracks, as depicted in Figure 1.3.2. Table 1.3.2 lists the chemical composition and Ms temperature of Fe–Ni-based LTT alloys.

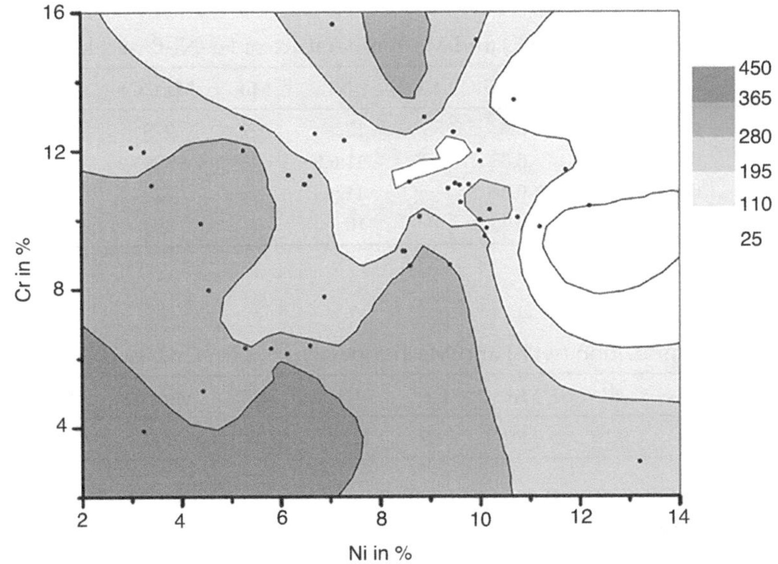

FIGURE 1.3.1
Ms temperature of selected Cr/Ni alloyed LTT fillers [17,26].

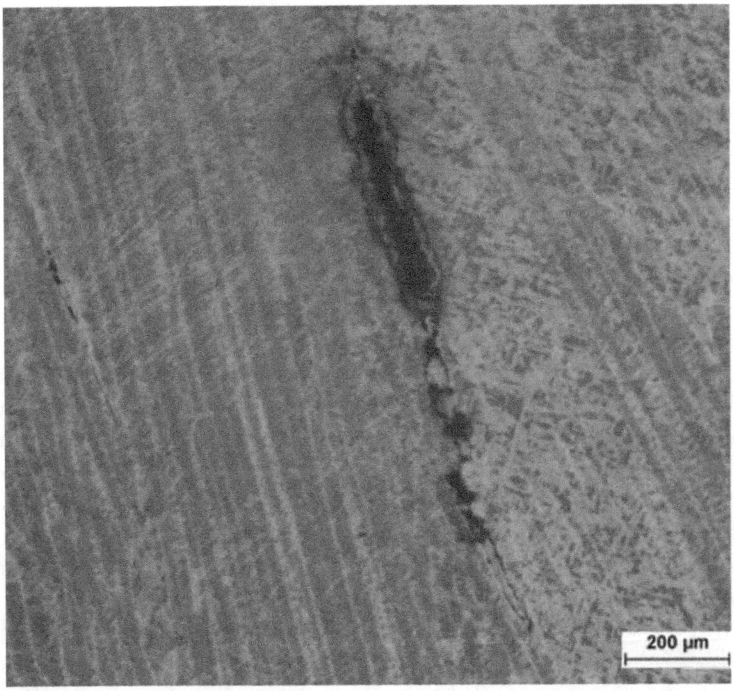

FIGURE 1.3.2
Solidification crack found in the 12%Ni LTT weld metal.

TABLE 1.3.2

Chemical Composition (wt%) and Ms Temperature of Fe–Ni-Based LTT Alloys

Name	C	Si	Mn	Cr	Ni	Mo	Ms (°C)	References
Series B	0.03	0.65	0.5	1	12	0.5	275	[20,27]
D	0.046		0.33	0.96	11.62		396	[28]
LTTW	0.04	0.32	0.36		11.9		420	[29]
LTT	0.06	0.25	0,17		10		350	[30]

TABLE 1.3.3

Chemical Composition (wt%) and Ms Temperature of Fe–Cr-Based LTT Alloys

Name	C	Si	Mn	Cr	Ni	Mo	Ms (°C)	References
A6	0.08	0.19	0.89	14.7	0.27	0.04	360	[23]

1.3.3 Fe–Cr-based Alloy

Although chromium is a ferrite stabilizer, adding about 10–16 wt% can help decrease the transformation temperature. Generally, the alloy can be easily designed to contain a minimum of 12% chromium to achieve "stainless" properties [5,16]. A low carbon content is also necessary to prevent the formation of chromium carbides during the reheating process. The formation of such precipitates could lead to intergranular attack due to chromium depletion. Additionally, chromium additions enhance the oxidation resistance of the alloy. Table 1.3.3 lists the chemical composition (wt%) and Ms temperature of Fe–Cr-based LTT alloys.

1.3.4 Fe–Cr–Mn-based Alloy

An alternative composition was tested, utilizing appropriate levels of manganese to substitute for nickel. The austenite-stabilizing effect of manganese is less pronounced than that of nickel; specifically, exceeding critical levels of the alloying element can result in incomplete martensitic transformation. At very high nickel contents (exceeding 10%), retained austenite remains stable even at room temperature. Manganese contents of approximately 14% can lead to the presence of retained austenite [17,31]. Martensitic Fe–Cr–Mn alloys can be very brittle [32,33], likely due to the co-segregation of manganese and phosphorus at prior austenite grain boundaries [33]. Research by Kannengiesser indicated that the 11%Cr–6%Mn LTT alloy was prone to root cracking in single-layer welds. However, employing multi-pass welding can prevent cracking by tempering and increasing

TABLE 1.3.4

Chemical Composition (wt%) and Ms Temperature of Fe–Cr–Mn-Based LTT Alloys

LTT Alloy	C	Si	Mn	Cr	Ni	Ms (°C)	References
C	0.085		6.38	10.99	0.02	137	[28]
Mn5	0.04	0.27	9.0	10.8	0.1	150	[31]

TABLE 1.3.5

Chemical Composition (wt%) and Ms Temperature of MX-4AD LTT Deposited Metal

Name	C	Si	Mn	Ms (°C)	References
MX-4AD	0.03	0.38	4.3	472	[34]

hydrogen effusion [28]. Table 1.3.4 lists the chemical composition and Ms temperature of Fe–Cr–Mn-based LTT alloys.

1.3.5 Fe–Mn-based Alloy

Miyata and Suzuki [34] conducted a study to replace Ni with Mn and developed the 4.3%Mn LTT welding wire (Kobelco: MX-4AD), suggesting that Mn can more effectively improve fatigue resistance than Ni. Furthermore, when used for additional beads, the 4.3%Mn LTT welding wire was confirmed to better improve fatigue resistance than other methods such as grinding treatment. Table 1.3.5 lists the chemical composition and Ms temperature of 4.3%Mn LTT deposited metal.

1.3.6 Fe–Mn–Ni-based Alloy

Miyata and Suzuki [34] slightly adjusted the alloy composition to meet practical toughness requirements, resulting in the development of the 3%Ni–3%Mn welding rod (Kobelco: LB-3AD). The authors fabricated fillet joints using the conventional welding wire (Kobelco: MX-200) and treated the weld toes with 4.3%Mn, 3%Ni–3%Mn, peening, and grinding, respectively. Subsequently, they studied and compared the three-point bending fatigue test results among fillet joints with weld toes treated using different methods. As shown in Figure 1.3.3, the joint treated with 3%Ni–3%Mn as additional weld beads could enhance the fatigue strength more effectively than improving weld toe fatigue resistance through grinding or peening. The fatigue enhancement is primarily attributed to the transverse compressive residual stress induced by LTT welding. Table 1.3.6 lists the chemical composition and Ms temperature of 3%Ni–3%Mn LTT deposited metal.

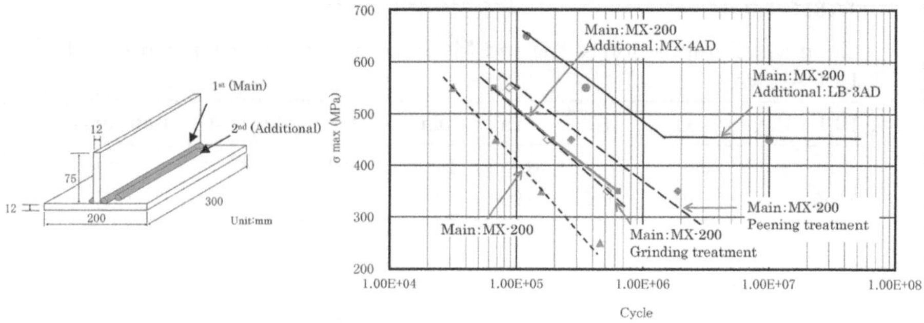

FIGURE 1.3.3
Comparison of bending fatigue strength of each treatment [34].

TABLE 1.3.6
Chemical Composition (wt%) and Ms Temperature of LB-3AD Deposited Metal

Name	C	Si	Mn	Ni	Ms (°C)	References
LB-3AD	0.03	0.41	3.3	3.3	408	[34]

1.4 Key Factors in LTT Welding Materials Design for Applications

Designing an LTT alloy is not a straightforward endeavor, as numerous factors need to be considered. When utilizing LTT alloys in multi-pass welding, a wide range of factors must be evaluated. These include the Ms temperature, residual stress, alloy compositions, dilution effect, mechanical properties, joint geometry, structural restraint (such as steel grade and plate dimensions), interpass temperature, layering method, type of loading, and susceptibility to hot, cold, and corrosion cracking in the weldment. We believe that these factors have posed significant obstacles to the economic commercialization of LTT alloys. It appears that, for the time being, the optimal approach is to tailor all the conditions in the design of the LTT alloy specifically for welding a particular type of steel and dimension.

1.4.1 Dilution Effect

Dilution can markedly modify the Ms temperature, which may potentially impede complete transformation. For example, when welding low-alloy steel, the element concentration in the welds was substantially lower than that of the initial pure alloy material, with dilution levels ranging from 25% to 35% for single-pass welds and approximately 10% for two-pass welds [35].

This study highlights the impact of single-pass dilution effects and the enhancements in fatigue performance that can be achieved. Feng et al. [36] demonstrated that the welding position and parameters can significantly influence dilution and the resulting Ms temperature range (150°C–250°C) when using the 16%Cr–8%Ni LTT alloy. This temperature range is optimal for generating compressive residual stress, thereby extending the fatigue life.

1.4.2 Mechanical Properties

When developing LTT welding fillers, the primary objective appears to be achieving a lower Ms temperature. However, this benefit is negated if other properties are neglected. In addition to considering dilution and weldability, it is crucial to ensure sufficient strength to join high strength steels. On the other hand, LTT weld metals with inadequate toughness may not be suitable for industrial applications. Therefore, a satisfactory balance among the aforementioned factors must be achieved when designing the alloy composition of LTT welding fillers.

Figure 1.4.1 shows the range of possible YS and tensile strength (TS) for some reported LTT alloys, as presented in Ref. [26]. The YS ranges from 300 to 1,100 MPa, and the TS ranges from 500 to 1,300 MPa. Increasing Cr and Ni lowers the Ms temperature; thus, reducing Ms temperature through the addition of Cr and Ni is accompanied by a reduction in YS, TS, and, consequently, impact toughness. As Ni stabilizes austenite (γ), the increase in the volume fraction of austenite (γ) with increasing Ni appears to be responsible for the reduction in mechanical properties.

Table 1.4.1 lists the toughness of LTT welds from various sources. Generally, the toughness of LTT weld metal is influenced by several interrelated factors, including the welding process, alloy composition, and microstructure. By comprehensively considering these factors, the toughness of LTT weld metal can be effectively improved to meet the requirements of various engineering applications.

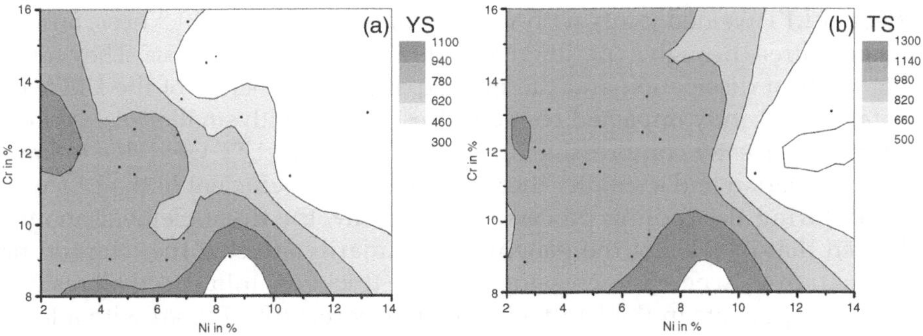

FIGURE 1.4.1
Effects of increasing Cr–Ni on YS and TS of welding filler [17,26].

TABLE 1.4.1

Mechanical Properties of LTT Weld Metals from Various Sources

				Toughness(J)				
LTT Alloy	**YS (MPa)**	**TS (MPa)**	**EL (%)**	**−40°C**	**−20°C**	**0°C**	**20°C**	**References**
10Cr–10Ni	822	1192	36		39			[37]
B206	700–850	950–1,050	>15	>100			>110	[21]
13Cr/LC35	680	1,050	15	30			36	[21]
A6	500	624				8	10	[2]
B5	886	1,002				12	14	[2]
C5	702	1,155				14	14	[2]
Alloy 2b	838	1,069	17.6		53		72	[5]
LTTE	1,135	1,287	6	17	15		20	[27]
Series B				22	27		28	[27]
LTT-C	736	1,127	13	49				[35]
Camalloy	838				53		72	[38]
LTTA	1,055	1,216	6.9		20			[39]
LTTB	254	990	9.3		28.1			[39]

1.4.3 Joint Geometry and Structural Restraint

1.4.3.1 Type of Joint

Research has confirmed that the optimal Ms temperature of the welding materials required for introducing compressive residual stress and minimizing distortion in a butt weld is approximately 200°C. In contrast, for a fillet-welded T-joint, the optimal Ms temperature is around 400°C, where transient angular distortion is minimized [16,40].

1.4.3.2 Size Effect

Feng et al. [41] investigated the relationship among residual stress, the size effect of LTT welded joints with varying plate width and thickness, and Ms temperatures through experimental and finite-element analysis. They found that the heat dissipation and thermal expansion coefficient of the LTT weld metal significantly impacted residual stress. Joints with smaller plate widths resulted in greater compressive residual stress in the LTT weld due to lower heat dissipation and a smaller thermal expansion coefficient of the LTT weld metal during the cooling process. Additionally, the finite-element analysis showed that increasing the plate width primarily affected the longitudinal residual stress, whereas increasing the plate thickness influenced all residual stress components in the LTT weld. Furthermore, an LTT weld with a lower Ms temperature of 191°C led to greater compressive residual stress and was less sensitive to the LTT joint size compared to an Ms temperature of 398°C.

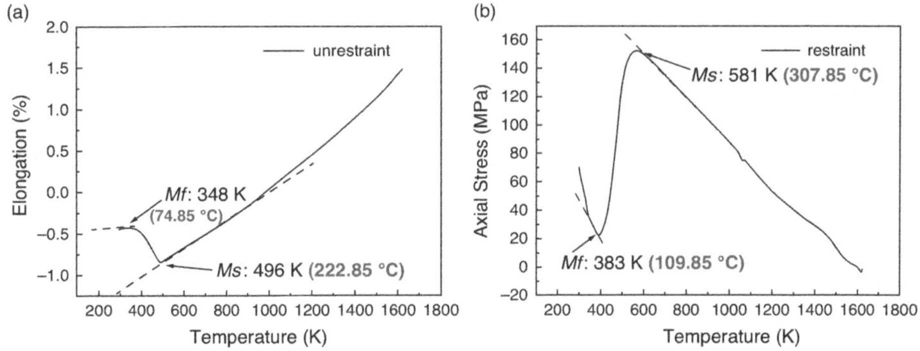

FIGURE 1.4.2
Effect of restraint state on martensitic transformation temperatures: (a) dilatometric curve of LTT weld metals under unrestrained conditions and (b) evolution of axial stress of LTT weld metals under restrained conditions [42].

1.4.3.3 Structural Restraint

Zhou et al. [42] discovered that the Ms temperature and martensitic transformation temperature range of the 11%Cr–5%Ni LTT weld metal under restrained conditions increased by 85°C and 50°C, respectively, compared to those under unrestrained conditions, as shown in Figure 1.4.2. This increase is attributed to the significant rise in Gibbs free energy produced by the restraint stress, which affects the martensitic transformation.

1.4.4 Interpass Temperature

The effect of interpass temperature on residual stress in the LTT multi-pass welds was investigated by Ramjaun et al. [43] and Feng et al. [44]. They found that when the interpass temperature exceeded the martensite start (Ms) temperature, the entire LTT multi-pass weld transformed as a single entity, thereby increasing the magnitude of compressive residual stress. Otherwise, the accumulation of thermally induced strains mitigates or eliminates the beneficial compressive stress state previously produced.

1.4.5 Type of Loading

LTT welds have shown a considerable enhancement in fatigue resistance compared to conventional joints in constant-amplitude fatigue testing, with some studies reporting an approximate 40% increase in mean fatigue strength [21,45]. However, in variable-amplitude fatigue testing, the observed improvement was less pronounced (12%) compared to that found during constant-amplitude loading [46]. This was primarily attributed to the relaxation of compressive residual stress under high-amplitude loading during variable-amplitude fatigue.

1.4.6 Cold and Hot Cracking Susceptibility of LTT Weld

Cold cracking is closely associated with the combined presence of three factors: diffusible hydrogen, a susceptible microstructure, and tensile residual stress [47]. Generally, such cracking occurs in the temperature range between –50°C and 150°C [8]. Although the use of LTT welding filler is beneficial for mitigating tensile residual stress, the martensitic microstructure of some LTT weld metals is sensitive to cold cracking. To minimize susceptibility to cold cracking, Zenitani et al. [22] and Kromm et al. [48] lowered the Ms temperature to increase the retained austenite content at room temperature. Their results indicated that excellent cold cracking resistance was attributed to the presence of compressive residual stress in the weld zone and the retained austenite, as shown in Table 1.4.2 and Figure 1.4.3. From a cracking perspective, the concentrated strain field at the propagating crack tip can result in the solid-state transformation of the adjacent retained austenite into martensite. This strain-induced martensitic transformation consumes strain energy. Simultaneously, volume expansion occurs, which closes the crack and retards crack propagation. Moreover, retained austenite also facilitates hydrogen entrapment [49].

A major issue with highly alloyed steels is hot cracking. Particularly, the prevention of solidification cracking through deliberate alloying strategies is a crucial factor. For these reasons, alloy design must involve more than

TABLE 1.4.2

Crack Ratios and Hardness Values Obtained with Various LTT Consumables in the Tekken Test [48]

LTT Alloy	Hardness HV10	Surface Crack Ratio in %	Section Crack Ratio in %
8%Ni	417	70	60
10%Ni	401	0	30
12%Ni	394	0	0

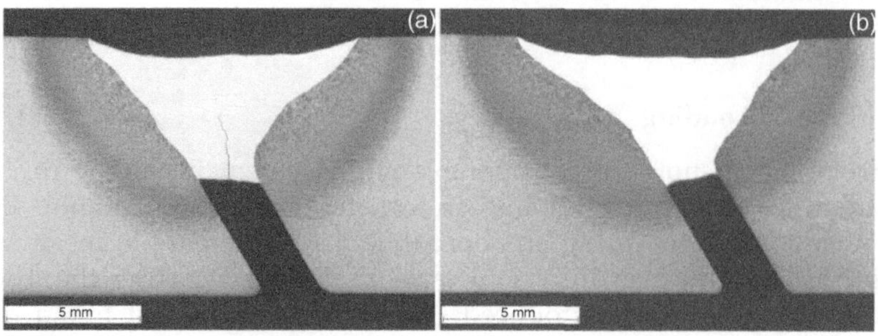

FIGURE 1.4.3
Cross sections of the joints produced with LTT consumables with varying Ni contents after the Tekken test: (a) 10%Ni and (b) 12%Ni [48].

simply adjusting the Ms temperature. Generally, adjusting the ratio of the Cr/Ni equivalent can influence the solidification mode and thus the susceptibility to hot cracking. The solidification mode can be classified as either ferritic or austenitic, depending on which phase the liquid metal first solidifies into. A ferritic–austenitic or austenitic–ferritic solidification mode indicates the leading phase during solidification, followed by the latter transformation into the trailing phase during the continuous cooling process [5]. If the liquid metal solidifies as δ-ferrite, which represents the ferritic–austenitic solidification mode, there is a greater resistance to hot cracking. Additionally, LTT weld metal with a lower percentage of discontinuous δ-ferrite retained at room temperature is considered beneficial for toughness. Meanwhile, a much finer austenite grain size can be achieved as it transforms from the decomposition of δ-ferrite, which is also advantageous for toughness [50]. However, an excessive amount of network-like δ-ferrite can be detrimental to impact toughness [51]. Furthermore, dilution by the base metal and the high-temperature cooling rate also affect the amount of δ-ferrite formed.

1.4.7 Corrosion Susceptibility of LTT Weld

A compositional mismatch of LTT alloys may lead to galvanic corrosion in some high strength steels in marine environments. Carbon (C) stabilizes austenite relative to ferrite and increases hardenability. However, high levels of carbon may lead to the formation of hard and brittle martensite upon cooling. In stainless steels, the formation of brittle carbides or chromium carbides can occur after welding, leading to increased susceptibility to corrosion and cracking. Additionally, impurity elements such as sulfur and phosphorus should be kept as low as possible, while elements such as molybdenum and tungsten can be added to prevent pitting corrosion. Meanwhile, the LTT alloy should be designed to contain a minimum of 12% chromium so that the alloy can be considered "stainless" [5,16].

References

1. A. Ohta, O. Watanabe, K. Matsuoka, C. Siga, S. Nishijima, Y. Maeda, N. Suzuki, and T. Kubo, Fatigue strength improvement by using newly developed low transformation temperature welding material. *Welding in the World*, 43 (1999), 38–42.
2. P.P. Darcis, et al., Cruciform fillet welded joint fatigue strength improvements by weld metal phase transformations. *Fatigue & Fracture of Engineering Materials and Structures*, 31(2) (2008), 125–136.
3. R. Gadallah, et al., Prediction of residual stresses induced by low transformation temperature weld wires and its validation using the contour method. *Marine Structures*, 44 (2015), 232–253.

4. A. Ohta, et al., Fatigue strength improvement of box welds by low transformation temperature welding wire and PWHT. *Welding in the World*, 2000(3) (2000), 44–45.
5. A. Shirzadi, et al., Stainless steel weld metal designed to mitigate residual stresses. *Science and Technology of Welding and Joining*, 14(6) (2009), 559–565.
6. F. Akyel, S. Olschok, and U. Reisgen, Reduction of distortion by using the low transformation temperature effect for high alloy steels in electron beam welding. *Welding in the World: Journal of the International Institute of Welding*, 65 (2021), 1.
7. Y. Nakanishi, *Welding distortion behavior of steel welds with low temperature transformation welding wire*. International Institute of Welding Japanese Delegation, IIW-Doc. XV-1196-05, Prague (2005).
8. G. Çam, O. Özdemir, and M. Koçak. Progress in low transformation temperature (LTT) filler wires. In *Proceedings of the 63rd Annual Assembly & International Conference of the International Institute of Welding*, İskenderun, Hatay and Pendik, İstanbul, Turkey (2010).
9. J. Hensel, et al., Capability of martensitic low transformation temperature welding consumables for increasing the fatigue strength of high strength steel joints. *Materials Testing*, 62(9) (2020), 891–899.
10. T. Nitschke-Pagel and H. Wohlfahrt, Residual stresses in welded joints – Sources and consequences. *Materials Science Forum*, 404–407 (2002), 215–226.
11. W.K.C. Jones and P.J. Alberry, A model for stress accumulation in steels during welding. *Metals Technology*, 11 (1977), 557–566.
12. K. Satoh, S. Matsui, and T. Machida, Thermal stresses developed in high-strength steels subjected to thermal cycles simulating weld heat-affected zone. *Journal of the Japan Welding Society*, 35(9) (1966), 780–789 (in Japanese).
13. H. Bühler and E. Scheil, Zusammenwirken von Wärme-und Umwandlungsspannungen in abgeschreckten Stählen. *Archiv für das Eisenhüttenwesen*, 6(7) (1933), 283–288.
14. H. Murata, N. Katoh, and H. Tamura, Effect of transformation on residual stress in welding-stress releasement by transformation superplasticity (part 5). *Quarterly Journal of the Japan Welding Society*, 11(4) (1993), 545–550 (in Japanese).
15. W. Wang, et al., New developed welding electrode for improving the fatigue strength of welded joints. *Journal of Material Science and Technology*, 18(6) (2002), 527–531.
16. S.W. Ooi, J.E. Garnham, and T.I. Ramjaun, Review: Low transformation temperature weld filler for tensile residual stress reduction. *Materials & Design* (1980–2015), 56 (2014), 773–781.
17. A. Kromm, J. Dixneit, and T. Kannengiesser, Residual stress engineering by low transformation temperature alloys—state of the art and recent developments. *Welding in the World*, 58(5) (2014), 729–741.
18. H.K.D.H. Bhadeshia, Developments in martensitic and bainitic steels: Role of the shape deformation. *Materials Science & Engineering A*, 378(1/2) (2004), 34–39.
19. H. Murata, N. Kato, and H. Tamura, Effect of alloying elements and transformation temperature on stress releasement: Stress releasement by transformation superplasticity (Part 2). *Transactions of the Japan Welding Society*, 23(2) (1992), 55–56.
20. J.A. Francis, et al. Transformation temperatures and welding residual stresses in ferritic steels. In *ASME Pressure Vessels & Piping Conference*, New York (2007).

21. J. Eckerlid, T. Nilsson, and L. Karlsson, Fatigue properties of longitudinal attachments welded using low transformation temperature filler. *Science and Technology of Welding and Joining*, 8(5) (2003), 353–359.
22. S. Zenitani, et al., Development of new low transformation temperature welding consumable to prevent cold cracking in high strength steel welds. *Science and Technology of Welding and Joining*, 12(6) (2007), 516–522.
23. M.C. Payares-Asprino, H. Katsumoto, and S. Liu, Effect of martensite start and finish temperature on residual stress development in structural steel welds. *Welding Journal*, 11 (2008), 87.
24. H. Lixing, et al., Ultrasonic peening and low transformation temperature electrodes used for improving the fatigue strength of welded joints. *Welding in the World*, 48(3–4) (2004), 34–39.
25. C. Miki and K. Anami, Improving fatigue strength by additional welding with low temperature transformation welding electrodes. *Steel Structure*, 1 (2001), 25–32.
26. A. Kromm, *Umwandlungsverhalten und Eigenspannungen beim Schweißen neuartiger LTT-Zusatzwerkstoffe*. Bundesanstalt für Materialforschung und-prüfung (BAM) (2011).
27. J. Francis, et al., The effects of filler metal transformation temperature on residual stresses in a high strength steel weld. *Journal of Pressure Vessel Technology*, 131 (2009), 4.
28. E. Kannengiesser, et al., Properties and weldability of modified low transformation temperature filler wires. *Welding Journal of the International Institute of Welding*, 1 (2015), 5587.
29. H. Murakawa, et al., Effect of low transformation temperature weld filler metal on welding residual stress. *Science and Technology of Welding and Joining*, 15(5) (2010), 393–399.
30. C. Miki, T. Hanji, and K. Tokunaga, Weld repair for fatigue-cracked joints in steel bridges by applying low temperature transformation welding wire. *Welding in the World*, 56 (2012), 40–50.
31. F. Martinez and S. Liu, *Development of Compressive Residual Stress in Structural Steel Weld Toes by Means of Weld Metal Phase Transformations*. ASM International, Georgia, USA (2005).
32. J. Bolton, E. Petty, and G. Allen, The mechanical properties of α-phase low-carbon Fe–Mn alloys. *Metallurgical Transactions*, 2 (1971), 2915–2923.
33. B. Edwards, M. Nasim, and E. Wilson, The nature of intergranular embrittlement in quenched Fe–Mn alloys. *Scripta Metallurgica*, 12(4) (1978), 377–380.
34. M. Miyata and R. Suzuki, Welding process and consumables aimed at improving fatigue strength of joints. *Kobe Steel Engineering Reports*, 65(1) (2015), 16–20 (in Japanese).
35. L. Karlsson and L. Mráz, Increasing fatigue life with low transformation temperature (LTT) welding consumables. *Zváranie Svarování*, 1 (2011), 8–15.
36. Z. Feng, et al., Development of 16Cr8Ni low transformation temperature welding material for optimal characteristics under various dilutions due to all repair welding positions. *Science and Technology of Welding and Joining*, 28(4) (2023), 305–313.
37. A. Ohta, et al., Fatigue strength improvement of lap welded joints by low transformation temperature welding wire—superior improvement with strength of steel. *Welding in the World*, 47(3) (2003), 38–43.

38. R.J. Moat, et al., Design of weld fillers for mitigation of residual stresses in ferritic and austenitic steel welds. *Science & Technology of Welding & Joining*, 16(3) (2011), 279–284.
39. S. Wu, et al., Mechanical properties of low-transformation-temperature weld metals after low-temperature postweld heat treatment. *Science and Technology of Welding and Joining*, 24(2) (2019), 112–120.
40. Y. Nakashima, et al., *Welding Distortion Behavior of Steel Welds with Low-Temperature Transformation Welding Wire*. IIW-Doc XV-1196-05 (2005).
41. Z. Feng, et al., Size effect on residual stress in low transformation temperature welded joints. *Marine Structures*, 78 (2021), 103001.
42. Z. Zhou, et al., Effect of restraint stress on martensite transformation in low transformation temperature weld metal. *Journal of Materials Science*, 55(5) (2020), 2202–2214.
43. T. Ramjaun, et al., Effect of interpass temperature on residual stresses in multi-pass welds produced using low transformation temperature filler alloy. *Science and Technology of Welding and Joining*, 19(1) (2014), 44–51.
44. Z. Feng, et al., Investigation of the residual stress in a multi-pass T-welded joint using low transformation temperature welding wire. *Materials*, 14(2) (2021), 325.
45. P.P. Darcis, et al., Cruciform fillet welded joint fatigue strength improvements by weld metal phase transformations. *Fatigue & Fracture of Engineering Materials & Structures*, 31(2) (2008), 125–136.
46. Z. Barsoum and M. Gustafson. Spectrum fatigue of high strength steel joints welded with low temperature transformation consumables. In *2nd International Conference on Fatigue Design*, Senlis, France, November 21–22 (2007).
47. E. Wilhelm, T. Mente, and M. Rhode, Waiting time before NDT of welded offshore steel grades under consideration of delayed hydrogen-assisted cracking. *Welding in the World*, 65(5) (2021), 947–959.
48. A. Kromm and T. Kannengießer, Characterizing phase transformations of different LTT alloys and their effect on residual stresses and cold cracking. *Welding in the World*, 55(3) (2011), 48–56.
49. M. Pinson, et al., The effect of quench cracks and retained austenite on the hydrogen trapping capacity of high carbon martensitic steels. *International Journal of Hydrogen Energy*, 46(29) (2021), 16141–16152.
50. Z. Zhang and R. Farrar, Columnar grain development in C–Mn–Ni low-alloy weld metals and the influence of nickel. *Journal of Materials Science*, 30(22) (1995), 5581–5588.
51. L. Karlsson, et al., Development of matching composition supermartensitic stainless steel welding consumables. *Svetsaren*, 3 (1999), 3–7.

2

Properties and Weldability of LTT Materials and Weld Metal

2.1 Properties of Low Transformation Temperature Materials and Weld Metals

2.1.1 Martensitic Transformation in Ni–Cr–Fe System

The martensitic transformation characteristics measured by the Transformation Measuring Apparatus (Fuji Electronic Industrial. Co.: formastor®) for the Ni–Cr–Fe alloy system with various chemical composition combinations are summarized in Table 2.1.1. This "formastor®" apparatus has the capacity to measure changes in temperature vs. time and transformation (thermal dilation) vs. time for all kinds of transformations of metallic materials.

In Table 2.1.1, chemical compositions, martensite start (Ms) temperature, and volume expansion strain $\Delta\varepsilon$ of martensitic transformation are listed for further analysis in the following items.

2.1.1.1 Estimated Equation of Ms Temperature

The Ms temperature for various low transformation temperature (LTT) materials with different chemical compositions is plotted on the Schaeffler diagram. In the Schaeffler diagram, the horizontal axis and vertical axis represent the Cr equivalent Cr_{eq} and Ni equivalent Ni_{eq}, respectively. The circle marks indicate the Cr_{eq} and Ni_{eq} positions for various LTT materials. The digital values near the circles in the diagram represent the Ms temperature. From the experimental data, Ms iso-value curves for 100°C, 200°C, 300°C, and 400°C are derived and plotted with bold lines. In this diagram, the microstructures of martensite, ferrite, and austenite are abbreviated as M, F, and A, respectively (Figure 2.1.1).

Each Ms iso-value curve has inflection points, making it difficult to predict the Ms temperature using a simple formula based on the chemical compositions. If the inflection points are connected, they correspond to the ratio of chemical compositions $Cr_{eq}/Ni_{eq}=0.5$ and 1.48. Then, with the aid of multiple regression analysis, the Ms temperature can be estimated and defined as Ms_{est} using the following equations in these three regions, respectively:

DOI: 10.1201/9781003601845-2

TABLE 2.1.1

Chemical Compositions and Martensitic Transformation Properties

Specimen No.	Chemical compositions														
	C (%)	Si (%)	Mn (%)	P (%)	S (%)	Ni (%)	Cr (%)	Mo (%)	Nb (%)	Al (%)	Ti (%)	Cr_{eq}	Ni_{eq}	Cr_{eq}/Ni_{eq}	Ms (Measured) (°C)	Expansion $\Delta\varepsilon$ (%)
1	0.040	0.22	0.63	0.010	0.004	9.5	12.5	0.26				13.1	11.0	1.19	–	–
2	0.063	0.42	1.44	0.011	0.001	9.2	11.4	0.22		0.022	0.021	12.3	11.8	1.04	–	0.100
3	0.048	0.38	1.21	0.015	0.002	7.1	11.9	0.23		0.022	0.064	12.7	9.2	1.39	184	0.466
4	0.072	0.35	1.38	0.015	0.004	6.8	12.9	0.10		0.011	0.001	13.5	9.7	1.40	150	0.442
5	0.069	0.39	1.13	0.008	0.002	7.7	9.8	0.11	0.004	0.012	0.020	10.5	10.3	1.02	171	0.546
6	0.064	0.48	1.39	0.016	0.005	10.7	7.5	0.01		0.026	0.300	8.2	13.3	0.62	133	0.475
7	0.085	0.44	1.61	0.018	0.003	6.8	12.9	0.12		0.010	0.011	13.7	10.2	1.35	124	0.438
8	0.062	0.40	1.30	0.016	0.003	7.2	12.8	0.10		0.012	0.072	13.5	9.7	1.39	146	0.454
9	0.052	0.27	0.96	0.006	0.002	9.6	10.3	0.11		0.014	0.002	10.8	11.6	0.93	126	0.480
10	0.050	0.30	1.00	0.005	0.002	9.6	11.0	0.15		0.010		11.6	11.6	1.00	110	–
11	0.048	0.26	0.57	0.020	0.010	9.5	10.5					10.9	11.2	0.98	170	–
12	0.030	0.35	0.68			9.7	11.2					11.7	10.9	1.07	149	–
13	0.066	0.42	1.40	0.011	0.002	7.6	9.0	0.24		0.023	0.026	9.9	10.3	0.96	183	0.530
14	0.053	0.33	0.92	0.004	0.001	7.7	9.2	0.23		0.025	0.002	9.9	9.8	1.01	232	0.383
15	0.060	0.34	0.94	0.005	0.004	7.1	7.8	0.12		0.019	0.001	8.4	9.4	0.90	250	0.355
16	0.076	0.42	1.20	0.009	0.002	7.2	9.6	0.09		0.015	0.015	10.3	10.1	1.02	183	0.449
17	0.074	0.38	1.17	0.008	0.002	6.9	8.9	0.10		0.014	0.020	9.6	9.7	0.99	192	0.471
18	0.061	0.54	1.50	0.015	0.006	7.4	8.5	0.12		0.004	0.034	9.4	10.0	0.94	218	0.551
19	0.063	0.47	1.49	0.015	0.005	6.6	8.3	0.10		0.003	0.030	9.1	9.2	0.99	237	0.530
20	0.070	0.42	1.53	0.015	0.006	6.4	7.5	0.10		0.003	0.033	8.3	9.3	0.89	237	0.515
21	0.050	0.30	1.00	0.005	0.002	7.4	7.8	0.15		0.010		8.4	9.4	0.89	260	–

(Continued)

TABLE 2.1.1 (*Continued*)

Chemical Compositions and Martensitic Transformation Properties

Specimen No.	Chemical compositions C (%)	Si (%)	Mn (%)	P (%)	S (%)	Ni (%)	Cr (%)	Mo (%)	Nb (%)	Al (%)	Ti (%)	Cr_{eq}	Ni_{eq}	Cr_{eq}/Ni_{eq}	Ms (Measured) (°C)	Expansion $\Delta\varepsilon$ (%)
22	0.054	0.30	0.78	0.003	0.001	6.0	6.5	0.27		0.029	0.002	7.3	8.0	0.90	308	0.425
23	0.061	0.38	1.29	0.010	0.002	5.9	6.8	0.29		0.027	0.027	7.6	8.3	0.92	253	0.433
24	0.080	0.35	1.46	0.015	0.005	5.9	6.9	0.10		0.003	0.028	7.5	9.1	0.83	255	0.484
25	0.064	0.36	1.00	0.006	0.004	6.4	7.0	0.12		0.019	0.005	7.7	8.8	0.87	255	0.418
26	0.080	0.39	1.12	0.008	0.006	5.3	5.8	0.12		0.020	0.012	6.5	8.2	0.79	285	0.421
27	0.072	0.41	1.16	0.008	0.006	4.7	4.9	0.14		0.017	0.021	5.6	7.4	0.75	340	0.292
28	0.055	0.26	0.82	0.005	0.008	8.1	5.1	0.08		0.013	0.001	5.5	10.2	0.54	262	0.386
29	0.083	0.42	1.25	0.010	0.006	3.6	3.4	0.14		0.017	0.028	4.2	6.7	0.62	373	0.225
30	0.020	0.34	0.44			8.2	16.1					16.6	9.0	1.84	29	–
31	0.056	0.38	1.31	0.015	0.001	6.2	12.6	0.25		0.020	0.002	13.4	8.5	1.57	136	0.498
32	0.055	0.37	1.31	0.015	0.001	5.8	12.1	0.23		0.017	0.008	12.9	8.1	1.59	181	0.419
33	0.048	0.38	0.98	0.015	0.002	5.3	14.1	0.18		0.016	0.002	14.8	7.2	2.05	168	0.492
34	0.061	0.36	1.28	0.017	0.002	5.8	15.2	0.09		0.007	0.004	15.8	8.3	1.91	52	0.330
35	0.059	0.32	1.02	0.015	0.002	5.4	16.0	0.07		0.008	0.002	16.6	7.7	2.15	76	0.253
36	0.047	0.52	1.67	0.015	0.010	6.3	15.0	0.26		0.009	0.084	16.0	8.5	1.88	91	0.413
37	0.065	0.48	1.51	0.015	0.010	5.4	13.3	0.21		0.008	0.070	14.2	8.1	1.75	156	0.506
38	0.066	0.45	1.52	0.015	0.009	5.4	13.2	0.21		0.005	0.051	14.1	8.1	1.73	172	0.473
39	0.066	0.36	1.41	0.015	0.009	4.7	11.4	0.20		0.005	0.046	12.1	7.3	1.65	207	0.495
40	0.049	0.76	1.43	0.013	0.009	10.8	0.0			0.027	0.400	1.2	13.0	0.09	325	0.383
41	0.048	0.73	1.39	0.012	0.008	14.4	0.0			0.036	0.550	1.1	16.5	0.07	258	0.515
42	0.052	0.22	0.78	0.004	0.012	8.1	4.0	0.07		0.011	0.001	4.4	10.0	0.43	298	0.346
43	0.060	0.13	0.70	0.005	0.020	8.1	0.0	0.02		0.010	0.001	0.2	10.2	0.02	328	0.177
44	0.042	0.73	1.33	0.011	0.008	18.5	0.0			0.045	0.630	1.1	20.4	0.06	185	0.593

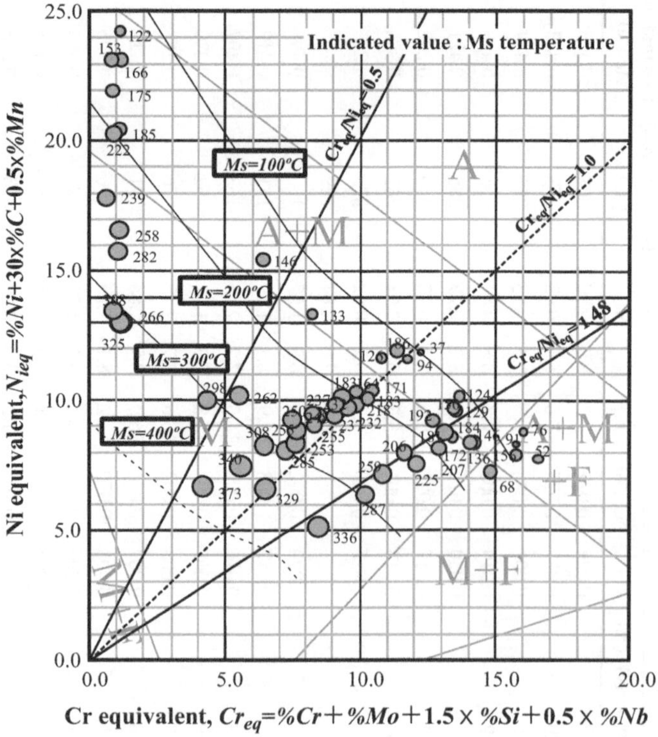

FIGURE 2.1.1
Relationship among Cr equivalent, Ni equivalent, and measured Ms temperature.

In region I ($Cr_{eq}/Ni_{eq} < 0.5$)

$$Ms_{est} = 535 - 14.6 \cdot Cr_{eq} - 16.3 \cdot Ni_{eq} \tag{2.1.1}$$

In region II ($0.5 \leqq Cr_{eq}/Ni_{eq} < 1.48$)

$$Ms_{est} = 630 - 17.8 \cdot Cr_{eq} - 25.5 \cdot Ni_{eq} \tag{2.1.2}$$

In region III ($1.48 \leqq Cr_{eq}/Ni_{eq} < 2.0$)

$$Ms_{est} = 901 - 35.1 \cdot Cr_{eq} - 31.6 \cdot Ni_{eq} \tag{2.1.3}$$

The correlation between measured Ms and estimated Ms_{est} in three regions is shown in Figure 2.1.2a–c, respectively. The estimation accuracy is very high and can be clearly observed.

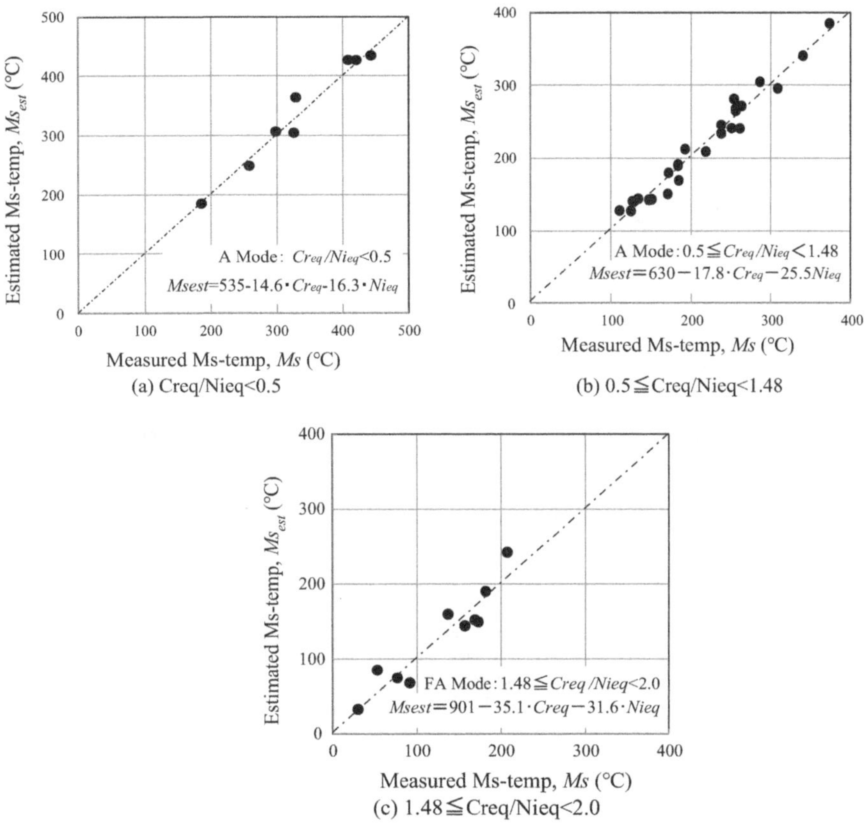

FIGURE 2.1.2
Comparison between estimated Ms_{est} and measured Ms temperature.

2.1.1.2 Martensitic Transformation Expansion

Figure 2.1.3 shows an example of the temperature–displacement relationship measured during the cooling from 1,200°C. The minimum value of the displacement (Disp[A]) when the temperature reaches the Ms is taken as the base value, and the difference ΔD between this and the displacement (Disp[B]) when the material is cooled down to room temperature (Disp[B])–(Disp[A]) is defined as the apparent martensitic transformation expansion.

If the martensitic transformation expansion ΔD is divided by the length of the specimen, the martensitic transformation strain $\Delta\varepsilon$ can be obtained. Figure 2.1.4 shows the influence of the Ms temperature on the expansion strain $\Delta\varepsilon$ [1]. When the Ms temperature is around 200°C, the expansion strain is largest and reaches about 0.5%. When the Ms temperature is between 100°C and 300°C, the expansion strain $\Delta\varepsilon$ is larger than 0.4%. Beyond this temperature range, it becomes smaller.

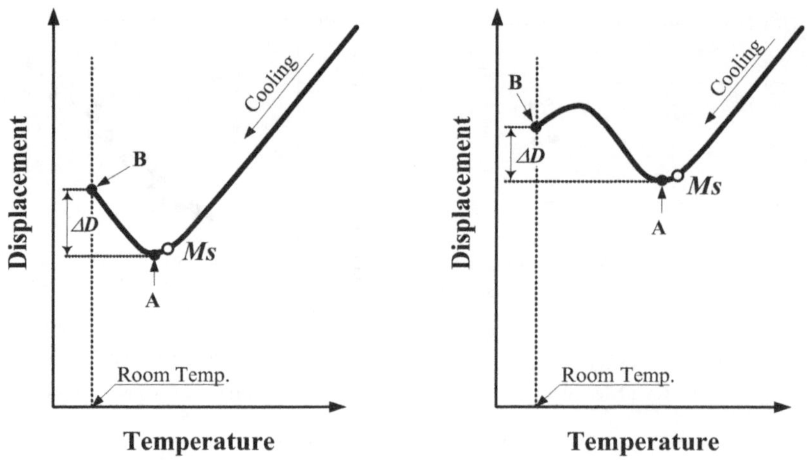

FIGURE 2.1.3
Definition of martensitic transformation expansion.

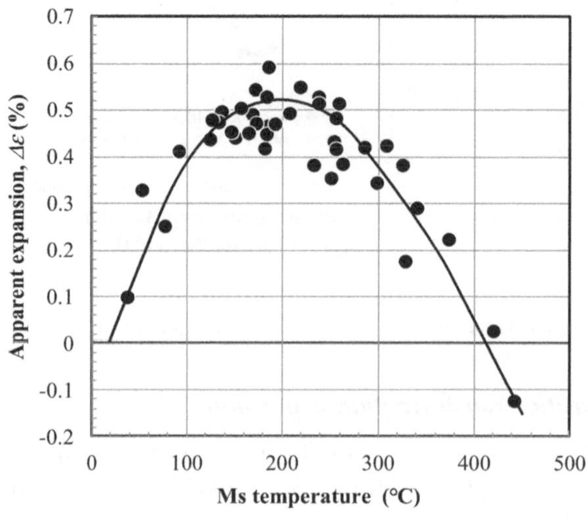

FIGURE 2.1.4
Relationship between Ms temperature and expansion strain.

2.1.1.3 Expression of Transformation Behavior

In the austenite–martensite transformation, martensite begins to form at a certain temperature denoted as the Ms temperature, and this transformation proceeds upon continuous cooling [2,3]. Magee [4] discussed the driving force for martensite nucleation from the viewpoint of thermodynamics and provided an expression for the volume fraction of martensite, f_m, as a function of temperature, T, as follows:

$$f_{\mathrm{m}} = 1 - \exp\left[\bar{V}\varphi\left(\partial \Delta G_y^{\delta-\alpha'}/dT\right)(\mathrm{Ms} - T)\right] \tag{2.1.4}$$

where $\bar{V}$ is the average volume of the newly formed martensite plates due to temperature variation dT, φ is a proportionality constant, and $\Delta G_y^{\delta-\alpha'}$ is the driving force (free energy) for the formation of martensite.

Koistinen and Marburger [5] measured the contents of martensite and retained austenite in carbon steels using X-ray techniques and proposed an empirical equation for f_{m} (actually identical to Eq. 2.1.4).

$$f_{\mathrm{m}} = 1 - \exp[-b(\mathrm{Ms} - T)] \tag{2.1.5}$$

where $b\left(= [df_{\mathrm{m}}/dT]_{T=Ms}\right) = 0.011$, and temperature is in °C. Following work by other researchers [2,6] shows that b is a material constant that varies with chemical composition. These results convincingly support Eq. (2.1.4) and demonstrate that Eq. (2.1.4) adequately describes the kinetic process of the austenite–martensite transformation in steels.

f_{m} in the cooling process can be measured using in situ equipment, such as in situ X-ray equipment. However, in addition to high costs, measurements are sometimes unavailable since such in situ equipment is not conventional. It is desirable to apply Eq. (2.1.5) to directly determine f_{m} at any temperature if b and Ms are known.

As the temperature decreases in the austenite region, the sample will shrink proportionally. The proportional coefficient is generally denoted as the expansion coefficient. We refer to this reduction in volume as the shrinkage caused by the expansion coefficient. When the temperature falls below Ms, austenite transforms into martensite, using the sample to expand because of the difference in volume between austenite and martensite. The newly formed martensite also shrinks as the temperature decreases, similar to austenite. Therefore, the total volume variation in the austenite–martensite transformation region depends on the sum of three variations: the shrinkage of the retained austenite, the shrinkage of the newly formed martensite, and the volume expansion using the austenite–martensite transformation). Apparently, the total volume change is directly correlated to f_{m}. If their relations can be expressed in equations, f_{m} can be deduced from the experimental data of dilatation.

Based on this idea, Yamamoto et al. [2] proposed a simple model to determine b and Ms from the measured dilation curve represented by the solid line in Figure 2.1.5.

Their model is briefly summarized by the following equations:

$$d_{i+1} = d_i - (f_{\mathrm{m}}\alpha_{\mathrm{m}} - f_\gamma \alpha_\gamma)\Delta T \tag{2.1.6}$$

$$E_{i+1} = d_{i+1} + \varepsilon_{\gamma-\mathrm{m}} \cdot f_{\mathrm{m}} \tag{2.1.7}$$

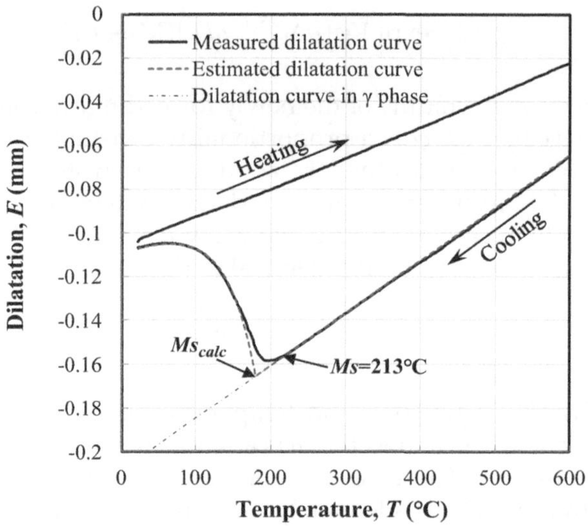

FIGURE 2.1.5
Measured dilatation curve using specimen after sub-zero treatment.

where i represents the i-th experimental point, d the shrinkage caused by the expansion coefficient, E the total dilatation at a given temperature, α_m the expansion coefficient of martensite, α_γ the expansion coefficient of austenite, f_m the volume fraction of martensite expressed with Eq. (2.1.5), f_γ the volume fraction of austenite, ΔT the temperature variation between any two temperatures, and $\varepsilon_{\gamma-m}$ the strain due to the lattice volume difference between austenite and martensite. Note that $f_m + f_\gamma = 1$. This implies that this model is only applicable to dual-phase austenite–martensite steels.

There are five unknown variables, b, Ms, α_m, α_γ, and $\varepsilon_{\gamma-m}$, in Eqs. (2.1.5–2.1.7), from which the Ms temperature can be estimated by fitting the measured curve, as shown in Figure 2.1.5.

Briefly, by submitting the initial values and the experimental α_m and α_γ into Eqs. (2.1.5–2.1.7), the value of E can be calculated. Then, by adjusting b, Ms_{calc}, and $\varepsilon_{\gamma-m}$ until the curve of experimental dilatation E *vs.* temperature is best fitted, the final fitted dilatation curve for the steels with a lower Ms temperature can be obtained and is shown by the broken line in Figure 2.1.5.

It is not easy to measure the martensite finish (Mf) temperature of materials at or near room temperature. Therefore, apart from the above measuring techniques, there is another approach to predict transformation characteristic parameters, such as the transformation expansion strain $\varepsilon_{\gamma-m}$ and the retained austenite f_γ.

Table 2.1.2 shows the predicted transformation expansion strain $\varepsilon_{\gamma-m}$ when the Cr content is fixed around 14% and the Ni content changes [3]. The transformation expansion strain tends to decrease with a decrease in the Ni equivalent. Referring to the Schaeffler diagram, this corresponds to the fact that

TABLE 2.1.2

Martensitic Transformation Parameters of Ni–Cr–Fe Weld

No.	C (%)	Ni (%)	Cr (%)	Cr_{eq} (%)	Ni_{eq} (%)	Ms_{calc} (°C)	Ms (°C)	$\varepsilon_{\gamma-m}$ (%)	Microstructure in Schaeffler Diagram
1	0.010	8.89	14.2	14.7	9.6	57	68	0.915	A+M
2	0.008	6.73	13.7	14.2	7.4	180	216	0.821	M
3	0.009	5.81	13.8	14.3	6.5	220	264	0.771	M+F
4	0.011	5.00	14.1	14.6	5.7	236	283	0.758	M+F
5	0.044	8.00	14.1	14.6	9.7	76	91	0.953	A+M
6	0.045	6.05	14.2	14.7	7.8	151	181	0.824	M+F
7	0.047	5.20	14.2	14.7	7.0	187	225	0.786	M+F
8	0.044	4.15	14.2	14.7	5.9	222	267	0.774	M+F

the lower the Ni equivalent, the more ferrite precipitates and the less austenite fraction transforms into martensite. For this reason, to get the maximum martensitic transformation expansion, the chemical compositions should be targeted in the region where ferrite does not precipitate.

When the same LTT welding wire is used to fabricate fillet, butt, and bead-on-plate welds, the welded metals have different dilution rates. The experimental dilatation curves of the welded metals with different dilution rates are shown in Figure 2.1.6. The transformation characteristic parameters are summarized in Table 2.1.3. The microstructures of these welded metals consist of martensite and austenite, with no ferrite present.

From Table 2.1.3, it can be observed that the martensitic transformation strain $\varepsilon_{\gamma-m}$ is nearly constant, and the *b* value decreases as the Ms temperature increases. Also, as shown in Figure 2.1.7, the retained austenite $f_\gamma(=1-f_m)$ at room temperature can be predicted based on the data obtaining from XRD [7] and saturation magnetization [8]. Corresponding to the retained austenite, the curve of transformation expansion strain $\Delta\varepsilon$ vs. Ms temperature can be plotted and is shown by a broken line in Figure 2.1.7. It can also be observed that the transformation expansion strain reaches its maximum value around the Ms temperature of 200°C.

2.1.2 Desirable Compositions of LTT Weld Metal

Although the purpose of LTT welding material is to generate compressive residual stress, it is also important to possess acceptable properties such as crack insusceptibility, ductility, and toughness.

2.1.2.1 Chemical Compositions for Introduction of Compressive Residual Stress

It has been described that compressive residual stress in welded metal can be generated through the expansion associated with martensitic transformation

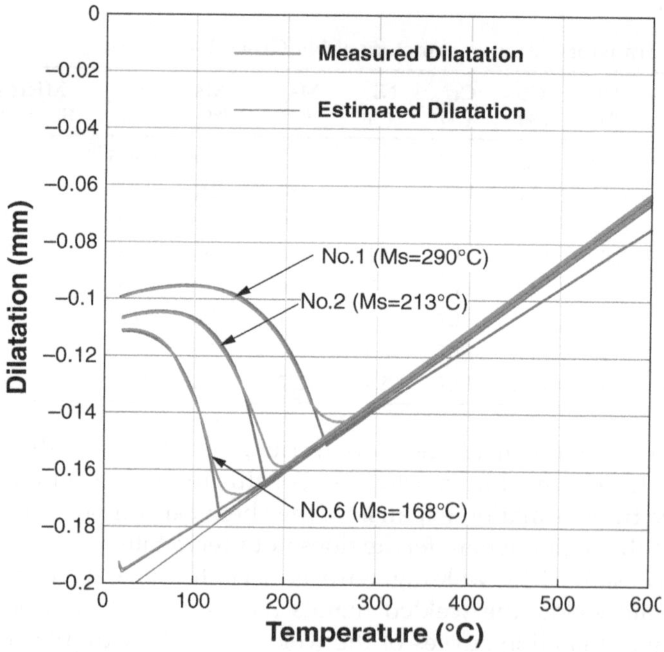

FIGURE 2.1.6
Samples of estimated dilatation curves and comparison with experiments.

TABLE 2.1.3
Transformation Parameters in Fitting Dilatation Curve

	Chemical Compositions						Transformation Parameter				
Specimen No.	C (%)	Ni (%)	Cr (%)	Cr_{eq} (%)	Ni_{eq} (%)	Ms (°C)	Ms_{calc} (°C)	B (1/°C)	$\varepsilon_{\gamma-m}$ (%)	f_g (%)	Remarks
1	0.068	5.2	10.1	10.6	7.6	290	245	0.015	0.90	3.1	Fillet
2	0.052	6.0	11.8	12.2	7.9	213	180	0.024	0.85	2.2	Butt
3	0.052	6.0	11.8	12.2	7.9	200	180	0.025	0.86	1.9	Butt
4	0.052	6.0	11.8	12.2	7.9	206	188	0.023	0.85	2.3	Butt
5	0.049	6.4	12.6	13.0	8.1	181	147	0.024	0.79	4.4	Overlay-1
6	0.044	6.6	12.9	13.4	8.2	168	130	0.028	0.87	5.3	Overlay-2
7	0.044	6.6	12.9	13.4	8.2	180	148	0.024	0.85	5.2	Overlay-2
8	0.043	6.6	13.0	13.5	8.2	170	140	0.022	0.84	7.2	Overlay-3

during cooling. Therefore, to achieve larger compressive residual stress, a greater transformation expansion is necessary.

The residual stress on the surface of a weld bead in a V-groove, measured using strain gauges via the stress release method, indicates that the Ms temperature significantly influences the magnitude of compressive residual stress, as shown in Figure 2.1.8 [1,9,10].

When Ms = 20°C (room temperature), the transformation expansion strain $\Delta\varepsilon$ is zero. This means that no martensitic transformation occurs up to room temperature and the residual stress is tensile, similar to that observed when using stainless welding wire.

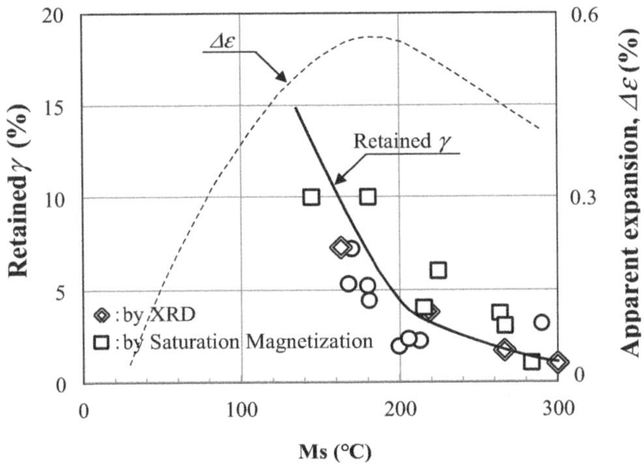

FIGURE 2.1.7
Correlation between retained γ and apparent martensitic expansion.

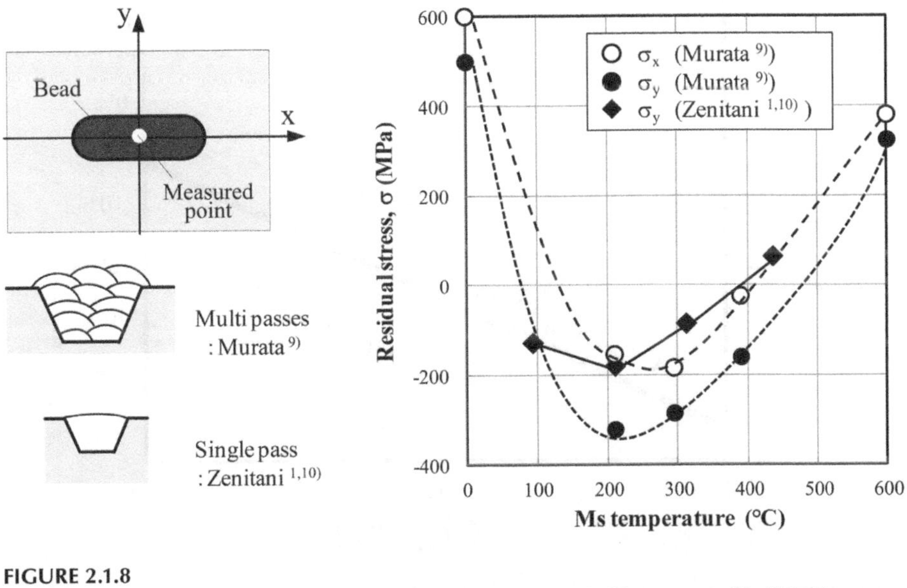

FIGURE 2.1.8
Relationship between residual stresses and Ms temperature in V-groove welds [1,9,10].

When Ms=400°C, the Mf temperature is around 200°C and the thermal stress at the Mf temperature is compressive. However, contraction due to subsequent cooling induces tensile stress again. As a result, the residual stress has a small magnitude, approaching zero. In other words, the transformation expansion can reduce the tensile welding residual stress.

When Ms=600°C, it behaves similarly to using conventional mild steel wire, resulting in tensile residual stress. The transformation expansion at such a high temperature has no influence on residual stress.

When Ms is 200°C, the maximum compressive residual stress is generated. Therefore, to reliably generate compressive residual stress, it is necessary to design chemical compositions such that the Ms temperature falls within the range of 100°C< Ms<350°C.

2.1.2.2 Chemical Compositions for Prevention of Solidification Cracking

In welding stainless steels, hot cracking is one of the major issues. Similarly, Ni–Cr–Fe-based LTT welding materials are also susceptible, particularly to solidification cracking in the weld metal.

It is known that material impurities with elements P and S increase solidification cracking susceptibility. Figure 2.1.9 illustrates the effect of Cr equivalent and Ni equivalent on solidification cracking of stainless steel weld metal when (P+S) contents are 0.05 mass% or higher [11]. Generally, when $Cr_{eq}/Ni_{eq} \leq 1.48$, susceptibility drastically increases [12].

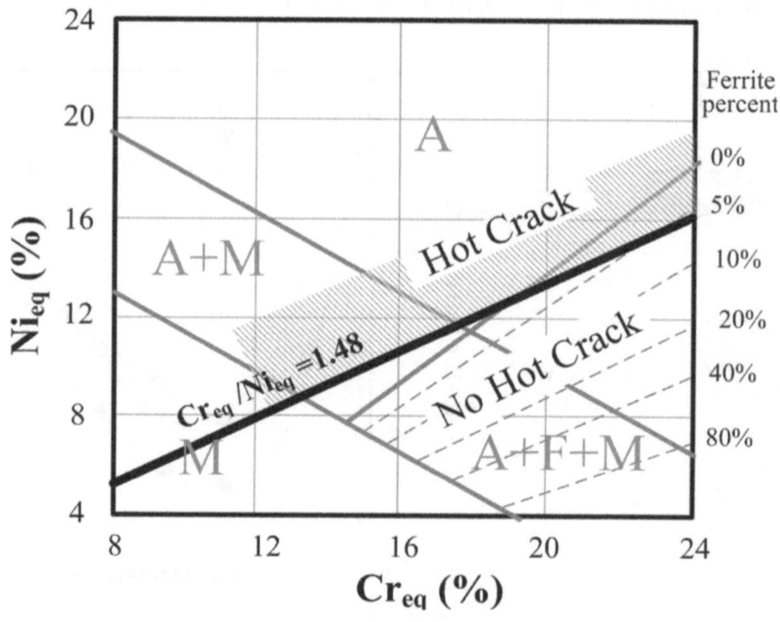

FIGURE 2.1.9
The effect of Cr equivalent and Ni equivalent on solidification cracking of stainless steel weld metal [11].

Figure 2.1.10 shows the relationship between microstructure and solidification mode [13].

When $Cr_{eq}/Ni_{eq} \geq 1.48$, the primary solidification phase is δ-ferrite (FA mode), which can reduce segregation of harmful elements (S, P) at grain boundaries due to higher solubility in the δ-phase. This greatly decreases cracking susceptibility.

When $Cr_{eq}/Ni_{eq} < 1.48$, the primary solidification phase is γ-austenite (AF or A mode), which leads to flatter grain boundaries and allows the liquid film to cover the entire grain boundary until the final solidification occurs, thereby increasing cracking susceptibility.

Thus, to mitigate hot cracking, it is essential to reduce P and S impurities and design compositions that favor FA mode. The optimal compositional range that satisfies both compressive residual stress generation and hot cracking resistance is indicated by the shaded region in Figure 2.1.11.

2.1.2.3 Chemical Compositions for Prevention of Cold Cracking

The main factors influencing welding cold cracking are (1) diffusible hydrogen content in the weld metal, (2) the existence of susceptible microstructures with higher hardness, and (3) tensile residual stress [14–19].

Therefore, reducing tensile residual stress and generating compressive stress via LTT welding could prevent cold cracking, as proposed by Zenitani et al. [1,10]. They conducted y-groove cracking tests (Tekken test) [20] using LTT welding covered electrodes and a conventional 800 MPa-class covered electrode, as shown in Figure 2.1.12.

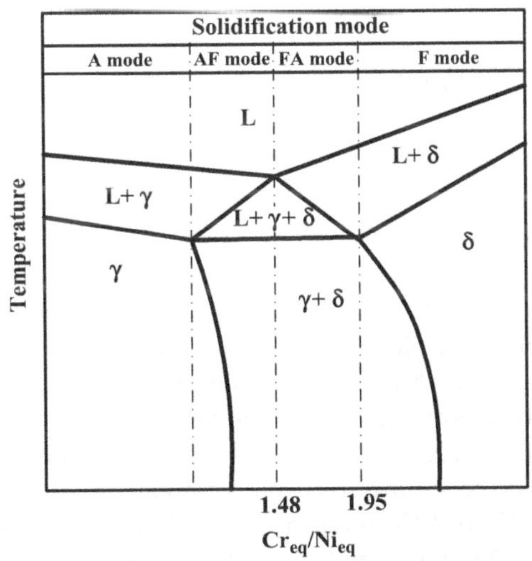

FIGURE 2.1.10
Solidification mode of welding stainless metal [13].

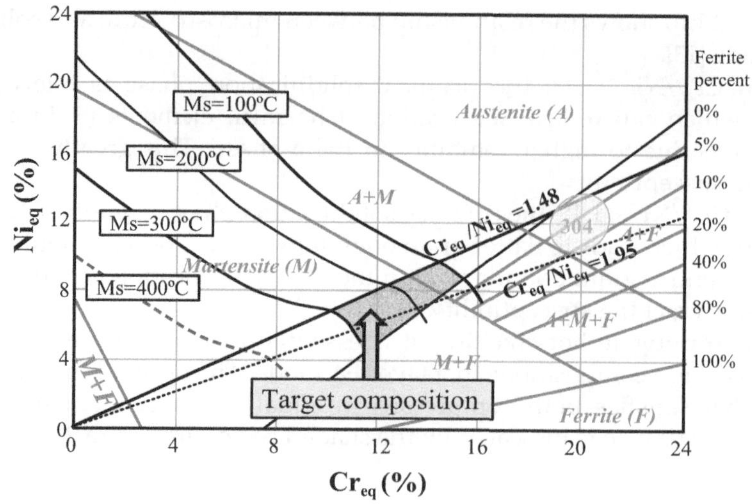

FIGURE 2.1.11
Desirable compositions of LTT weld metal to introduce compressive residual stress without hot crack.

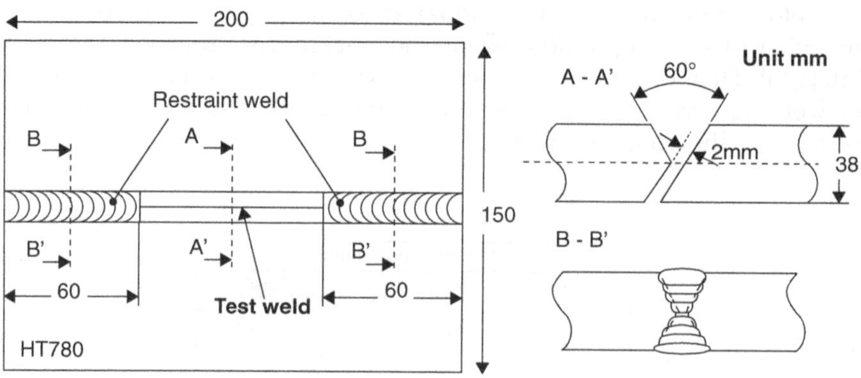

FIGURE 2.1.12
Specimen for y-groove weld cracking test (Tekken test).

In the cold cracking test, all welding rods were dried at 350°C for 2 hours before welding, and the welding was performed at room temperature (30°C) in an environment with 90% humidity, without preheating or post-heating.

Table 2.1.4 lists the chemical compositions, hardness (HV), and Ms temperature of different LTT weld metals in cracking tests. Figure 2.1.13 shows the relationship between Ms temperature and crack ratio. For higher Ms welds of LTT B (Ms=312°C) and Conventional D (Ms=436°C), the cracking ratio is 100%. For the lower Ms weld metal of LTT A (Ms=210°C), the crack ratio is dramatically reduced to 7%–29%.

TABLE 2.1.4

Chemical Compositions, HV, and Ms Temperature of Weld Metals

Weld Metal		C	Si	Mn	P	S	Mo	Ni	Cr	Ni_{eq} (%)	Cr_{eq} (%)	HV	P_{CM}	Ms (°C)
		Chemical Compositions (%)												
LTT	A	0.051	0.25	0.73	0.008	0.003	0.30	8.8	10.1	10.70	10.78	366	0.7	210
LTT	B	0.065	0.28	0.55	0.014	0.003	0.40	4.4	9.9	6.63	10.72	386	0.7	312
LTT	C	0.047	0.40	1.50	0.012	0.007	0.30	8.6	11.1	10.76	12.00	313	0.85	94
Conventional	D	0.091	0.52	1.18	0.015	0.006	0.31	1.76	0.6	5.08	1.69	334	0.25	436

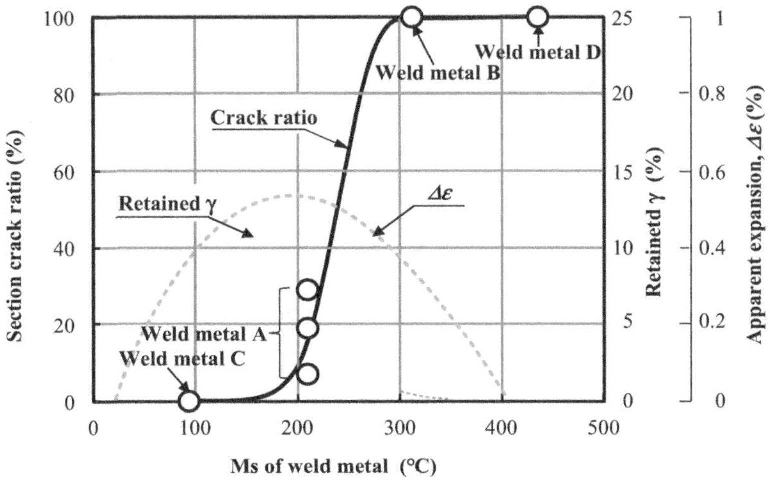

FIGURE 2.1.13
Relationship between Ms temperature of weld metal and crack ratio in y-groove cracking test.

TABLE 2.1.5

Diffusible Hydrogen Contents of Weld Metals in the Weld Cracking Test

		Diffusible Hydrogen Content (mL/100 g)			
Weld Metal		**1**	**2**	**3**	**Ave.**
LTT	A	5.07	5.36	5.56	5.33
LTT	B	6.45	5.12	5.59	5.72
Conventional	D	5.94	6.11	5.96	6.00

To explain the results in weld metals A, B, and D, the amount of diffusible hydrogen was measured in all the weld metals in accordance with JIS Z 3118, as shown in Table 2.1.5. Because the amount of diffusible hydrogen per 100 g of weld metal is 5–6 mL for all the weld metals, it is found that the significant decrease in the cracking ratio in weld metal A is independent of the amount of diffusible hydrogen.

The hardness of weld metals A and B is higher than that of weld metal D. This is because both weld metals A and B have a full martensite microstructure, while the microstructure of weld metal D consists of martensite and bainite. By comparing the microstructures of these weld metals, it can be observed that the microstructure has a limited influence on the cracking ratio.

From the above comparison and analysis, it is estimated that the contribution of compressive residual stress to the lower cracking ratio is significant, which is strongly influenced by the largest apparent expansion $\Delta\varepsilon$.

Although weld metal A has a lower Ms temperature and a small amount of hydrogen, cracks still occurred at a lower ratio. Therefore, controlling microstructures is still necessary to completely prevent cold cracks.

From the perspective of diffusible hydrogen content, austenite has a higher hydrogen solubility and a lower hydrogen diffusion coefficient compared to martensite. As a result, the diffusible hydrogen content decreases, and the accumulation of diffusible hydrogen in stress concentration zones (such as the root region) is alleviated, which may effectively prevent cold cracking [21,22]. Based on this concept, LTT C was proposed. Then, the cold crack test for LTT weld metal C was conducted and exhibited 0% cracking despite lower compressive stress than LTT weld metal A. Specifically, LTT weld metal C retained 15% austenite, suggesting that retained austenite also contributes to cracking resistance.

Figure 2.1.14 shows the measurement results of released hydrogen (diffusible hydrogen) over time after welding in specimens with varying amounts

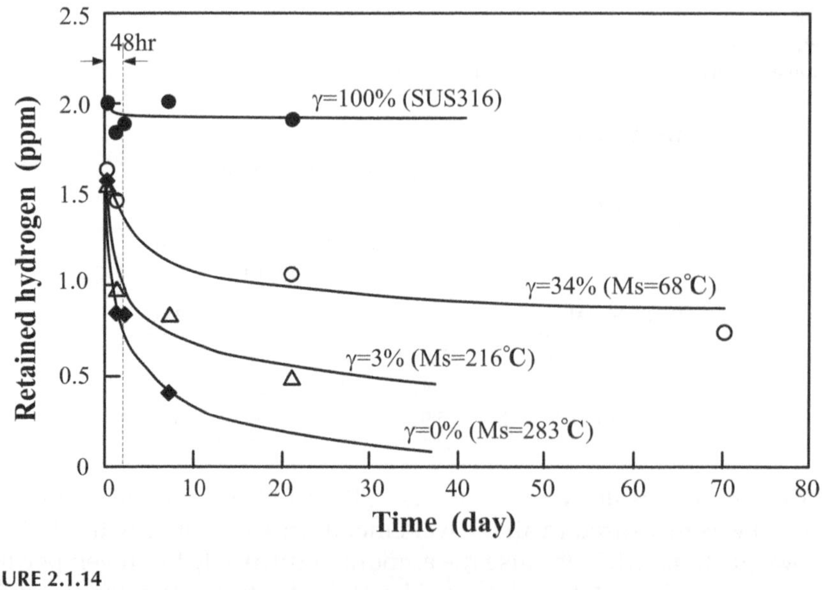

FIGURE 2.1.14
Relationship between hydrogen emission rate and retained γ in TDS measurement.

of retained γ (austenite), obtained using thermal desorption spectrometry (manufactured by ESCO) [8]. As the retained γ content increases, the hydrogen release rate slows, and a greater amount of hydrogen remains in the specimen over an extended period.

This can be understood as hydrogen being dissolved and trapped within the retained γ, leading to a reduction in diffusible hydrogen content proportional to the trapped amount. Furthermore, it demonstrates that even a few percent of retained γ can provide a trapping effect. In other words, based on the retained γ content shown in Figure 2.1.13, a trapping effect can be expected in weld metal compositions with an Ms point below 300°C.

Additionally, the zero cracking rate observed in FCW (flux-cored wire) welding in Figure 2.1.13 is likely due to the initial hydrogen content being lower than that in shielded metal arc welding.

The hydrogen-trapping effect of retained γ has been numerically analyzed in simulations of cold cracking tests [23,24]. Figure 2.1.15 presents the macro-analysis results of hydrogen diffusion and accumulation behavior after thermo-elasto-plastic analysis in a Y-slit model. The results show

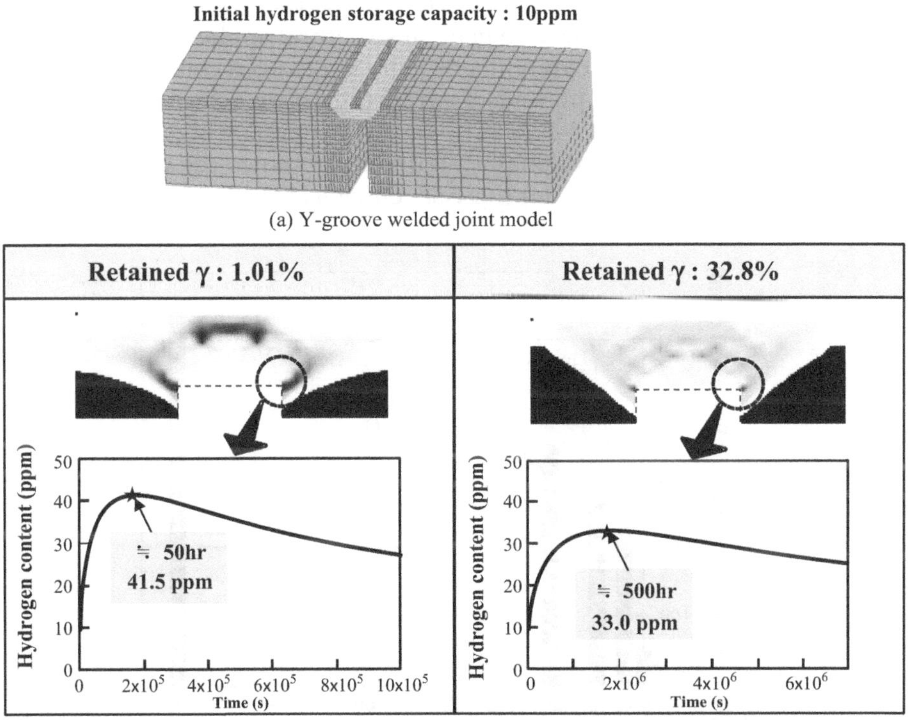

(b) Distribution of accumulation hydrogen around slit region of Y-groove welded joint.

FIGURE 2.1.15
Numerical analysis of diffusible hydrogen content at weld root.

that hydrogen accumulates at the slit corner (stress concentration zone) over time after welding at room temperature. Moreover, as the retained γ content increases, hydrogen accumulation in the stress concentration zone is mitigated. Conversely, when retained γ decreases, hydrogen accumulation occurs earlier, reaching higher maximum levels, creating conditions more prone to cold cracking.

Microscopic analysis confirms that the hydrogen content in the austenite phase is higher than in the martensite phase, reproducing the trapping effect.

Based on these findings, while reducing tensile residual stress or introducing compressive residual stress is effective in suppressing cold cracking, using retained austenite can further enhance this effect.

2.1.3 Improvement of Toughness in LTT Weld Metal

2.1.3.1 Effect of Oxygen Content on Toughness in LTT Weld Metal

The LTT weld metal has a significant advantage in reducing tensile residual stresses or even inducing compressive residual stresses in welded joints. However, this relies on the martensitic transformation, and the LTT weld metal primarily consists of a martensitic microstructure, which raises concerns regarding low toughness in general structural applications.

Notch toughness is strongly influenced by the oxygen content in the weld metal [25,26]. Figure 2.1.16 presents the measured oxygen content in the weld metal when using a 10%Cr–10%Ni LTT weld material with varying shielding gas compositions. Additionally, Figure 2.1.17 shows the Charpy impact absorbed energy (*vET*) at 0°C for these specimens [27].

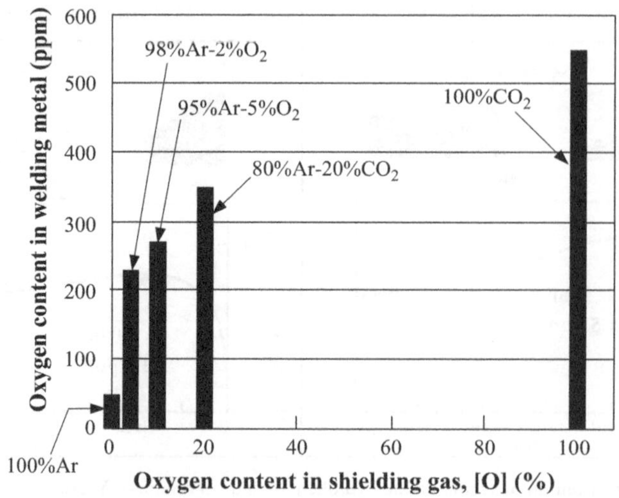

FIGURE 2.1.16
Effect of shielding gas composition on oxygen content in gas metal arc welding.

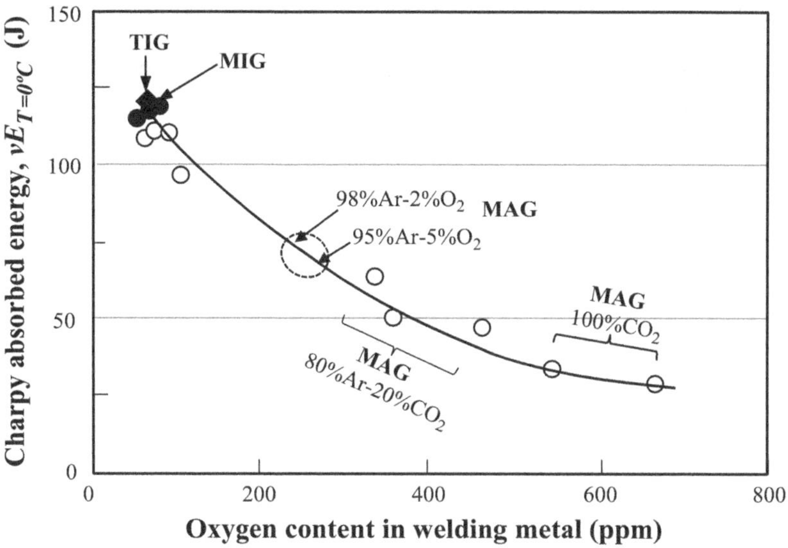

FIGURE 2.1.17
Relationship between oxygen content and Charpy absorbed energy in LTT weld metal.

In TIG and MIG welding with pure Ar shielding gas, the oxygen content in the weld metal is around 50 ppm, and the *vET* at 0°C is approximately 120 J. In contrast, MAG welding with 100% CO_2 shielding gas results in an oxygen content of about 600 ppm, causing the *vET* at 0°C to drop sharply to 30 J.

However, pure Ar-shielded MIG welding is rarely used in practical steel welding due to arc instability. Instead, MAG welding with Ar mixed with active gases (O_2 or CO_2) or 100% CO_2 shielding gas is typically employed.

MAG welding with Ar and 2%–5% O_2 shielding gas can improve the toughness *vET* at 0°C of LTT weld metals to approximately 70 J. Recent advancements have enabled stable MIG welding even with 100% Ar shielding gas [28,29], raising expectations for high toughness (*vET* > 100 J).

2.1.3.2 Effect of Retained Austenite on Fracture Toughness in LTT Weld Metal

Besides reducing oxygen content, another approach to enhancing the toughness of martensite-dominated LTT weld metals is utilizing the transformation-induced plasticity effect of retained austenite.

Figure 2.1.18 illustrates the relationship between the V-notch Charpy absorbed energy and retained γ (austenite) content in LTT weld metals produced by MIG welding with 100% Ar shielding gas. The results confirm that retained γ increases impact energy [30].

CTOD (Crack tip opening displacement) tests were conducted on 14%Cr–9%Ni and 14%Cr–7%Ni LTT weld metals (oxygen content <50 ppm) at

25°C, –80°C, and –196°C (Figure 2.1.19). The retained γ contents in these weld metals were 15% and 0%, respectively.

In the temperature range from room temperature to –80°C, the average CTOD value was 0.26 mm for the fully martensitic phase, while it increased to 0.35 mm with 15% retained austenite. At –196°C, the effect of retained austenite became even more pronounced, with the CTOD value reaching about 10 times that of the fully martensitic structure [30].

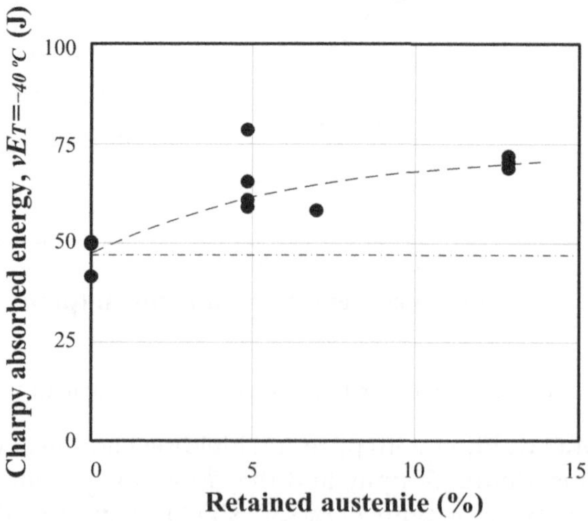

FIGURE 2.1.18
Relationship between retained austenite and Charpy absorbed energy in LTT weld metal.

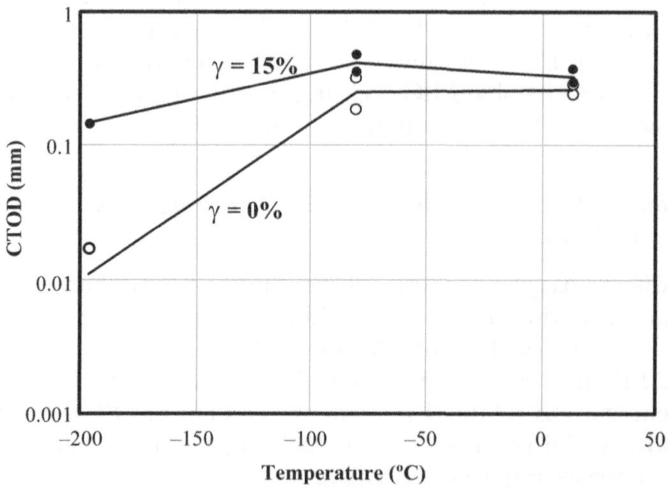

FIGURE 2.1.19
Effect of retained austenite content on critical CTOD.

2.1.3.3 Effect of Retained Austenite on Fatigue Crack Propagation in LTT Weld Metal

In Section 2.1.3.2, it was confirmed that retained austenite improves CTOD values. It can be hypothesized that retained austenite also retards fatigue crack propagation.

To investigate its effect, the fatigue crack growth behavior was monitored using the field signature method [31]. The results for two types of LTT weld metals (oxygen content <50 ppm), LTTW1 (γ=0%) and LTTW2 (γ=15%), are shown in Figure 2.1.20b [32]. Figure 2.1.20a illustrates the test specimen geometry, featuring a through-thickness notch machined at the center of the weld metal in a MIG welded butt joint. Here, the FC value is a parameter representing crack propagation [33], where an increase in the FC value indicates fatigue crack growth.

The results demonstrate that the weld metal containing 15% retained austenite exhibits approximately 2.5 times slower crack propagation compared to a fully martensitic single-phase weld metal (with no retained austenite).

The transformation-induced plasticity effect of retained austenite on fracture toughness and fatigue crack behavior, as discussed in Sections 2.1.3.2 and 2.1.3.3, appears to be particularly pronounced when the weld metal has a low oxygen content and few oxide inclusions.

Therefore, in practical applications, regardless of the welding processes (TIG, MIG, MAG), joint types (butt, fillet, overlay), or environment (shop, field), if welding can be performed in a low oxygen atmosphere to the greatest extent possible, a martensite-dominated LTT weld metal may not exhibit low toughness.

2.1.4 Corrosion Characteristics of LTT Weld Metal

LTT weld metal is a high-alloy weld metal containing a certain percentage of chromium (Cr) and nickel (Ni) to lower the martensitic transformation

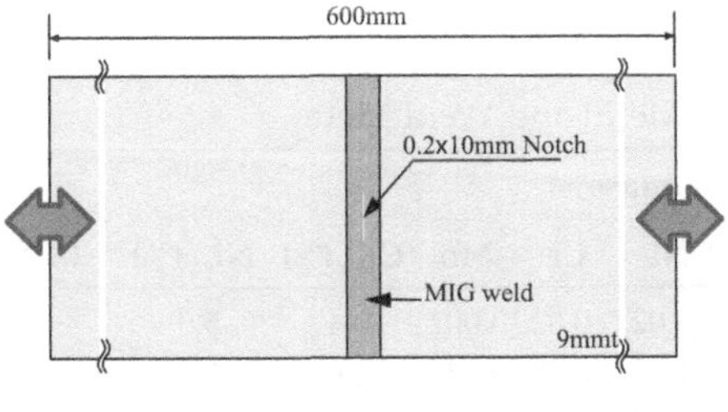

(a) Dimension of Specimen

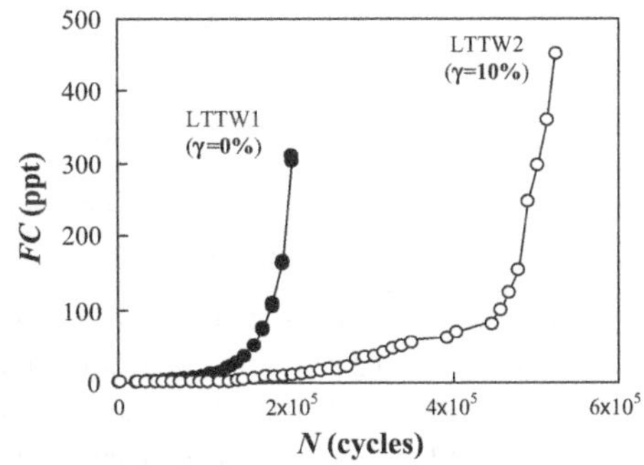

(b) Crack propagation behavior monitored by FSM

FIGURE 2.1.20
Effect of retained austenite content on fatigue crack propagation (ppt: pair per thousand).

temperature. When applied to high strength steel, it forms a dissimilar joint with different chemical compositions from both the base metal and the weld metal. Therefore, in outdoor structures where welded joints are exposed to rain and other severe environments, galvanic corrosion is a concern. Ohta et al. [34] and Zenitani et al. [35] conducted salt spray tests, atmospheric exposure tests, and cyclic dry/wet accelerated corrosion tests on high strength steel welded joints using LTT welding materials.

2.1.4.1 Atmospheric Corrosion Behavior of LTT Welded Joint

Zenitani et al. [35] prepared a joint with a Ni–Cr-based weld metal (Table 2.1.6) by machining a U-shaped groove (3 mm deep, 15 mm wide, with rounded edges of R7mm) on the surface of a 12 mm-thick high strength steel (SM490) and performing TIG welding using filler wire. The corrosion test specimens were welded joints where the weld reinforcement was removed by machining to smooth the surface. Atmospheric exposure tests were conducted to simulate conditions for bridges. The exposure site was the Choshi Testing Station of the Japan Weathering Test Center, with an average temperature of 14.8°C, average relative humidity of 78%, and average airborne salt deposition of 0.25 mdd (mg/100 cm^2/day).

An example of the specimen appearance after one year of atmospheric exposure is shown in Figure 2.1.21. Subsequently, rust was removed by acid cleaning, and the surface profile was measured using a laser microscope to evaluate corrosion. The results are shown in Figure 2.1.22a. The base metal was uniformly corroded in all specimens, with no significant difference observed; however, joints with higher Cr content exhibited slightly smoother surface roughness. Since the smoothed Ni–Cr weld metal surface remained uncorroded, the initial surface height of the base metal was taken as the reference.

The surface height profiles are shown in Figure 2.1.22. The marking lines near the fusion boundary shown in Figure 2.1.22a were measured, and the

TABLE 2.1.6

Chemical Composition and Ms Point of Base Metal and Weld Metal

		Chemical Composition (mass%)										
Material	No.	C	Si	Mn	P	S	Ni	Cr	Mo	Cr_{eq} (%)	Ni_{eq} (%)	Ms (%) (°C)
Base metal		0.16	0.24	1.41	0.013	0.009	0.02	0.03	0.002	0.4	5.5	
Weld metal	F9	0.08	0.4	1.59	0.001	0.005	1.89	0.43	0.53	1.6	5.1	422
	F5	0.08	0.42	1.25	0.01	0.006	3.57	3.41	0.14	4.2	6.6	366
	F1	0.06	0.34	0.94	0.005	0.004	7.09	7.81	0.12	8.4	9.4	254
	F17	0.06	0.45	1.6	0.013	0.003	9.73	11.24	0.12	12.0	12.3	<50
	F19	0.07	0.5	1.54	0.014	0.003	10.06	13.29	0.11	14.2	12.9	<50
	F20	0.07	0.42	1.59	0.014	0.002	14.87	19.22	0.08	19.9	17.8	<50

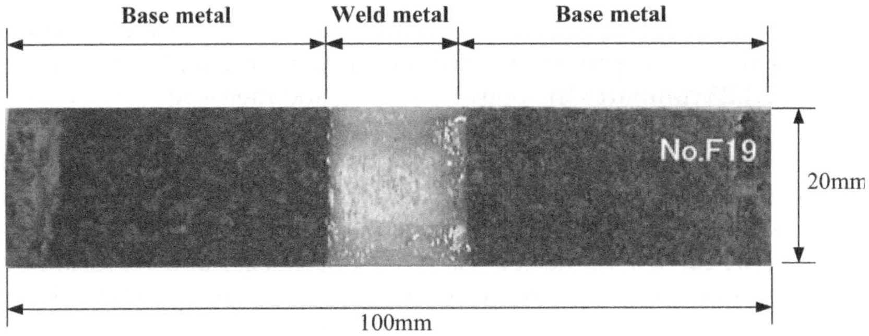

FIGURE 2.1.21
Specimen surface after outdoor exposure test.

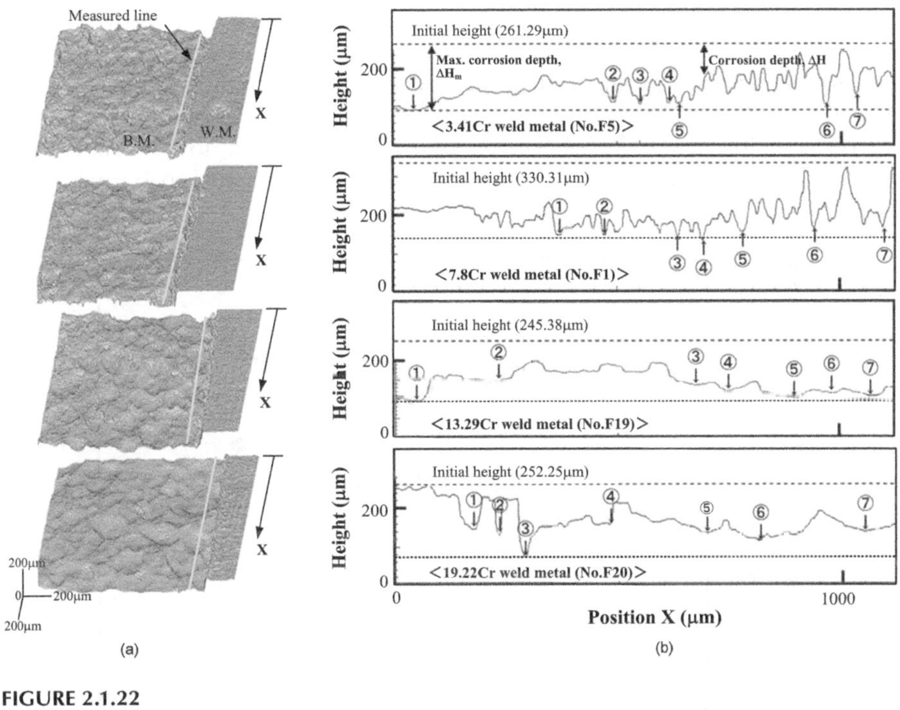

FIGURE 2.1.22
Three-dimensional surface profile near weld interface and corrosion depth distribution at weld interface metal.

results are shown in Figure 2.1.22b. The depth from the initial surface to the corroded surface was defined as corrosion depth (ΔH), and the maximum corrosion depth in each profile was denoted as ΔH_m.

The corrosion depths (ΔH) of localized pits (①–⑦) along each profile line in Figure 2.1.22b were plotted on Gumbel probability paper for extreme value statistical analysis to predict the maximum corrosion depth (ΔH_{me}).

Figure 2.1.23 compares the predicted ΔH_{me} and measured ΔH_m values with respect to the Cr content in the weld metal. The difference between the highest and lowest predicted ΔH_{me} values was approximately 47 μm. Similarly, the measured ΔH_m values showed no clear correlation with Cr content, as corrosion depth varied without a consistent trend.

In galvanic corrosion, the corrosion current density and corrosion rate are typically higher near the interface between dissimilar metals. Therefore, the potential for localized accelerated corrosion at the boundary between high-Cr–Ni weld metal and the base metal was investigated. Figure 2.1.24 shows the height profiles measured at varying distances from the fusion boundary (parallel to the fusion line) for a specimen (No. F19) with 13.3% Cr and 10.1% Ni weld metal. As before, extreme value statistical analysis was performed on the pit depths (①–⑦) along each profile line in Figure 2.1.24, and the predicted ΔH_{me} and measured ΔH_m are presented in Figure 2.1.25.

No significant difference in ΔH_{me} was observed based on measurement location, confirming that no localized accelerated corrosion occurred at the boundary between the high-Cr–Ni weld metal and the base metal.

In conclusion, the atmospheric exposure tests revealed no noticeable effect of Cr content in the weld metal or distance from the fusion boundary on

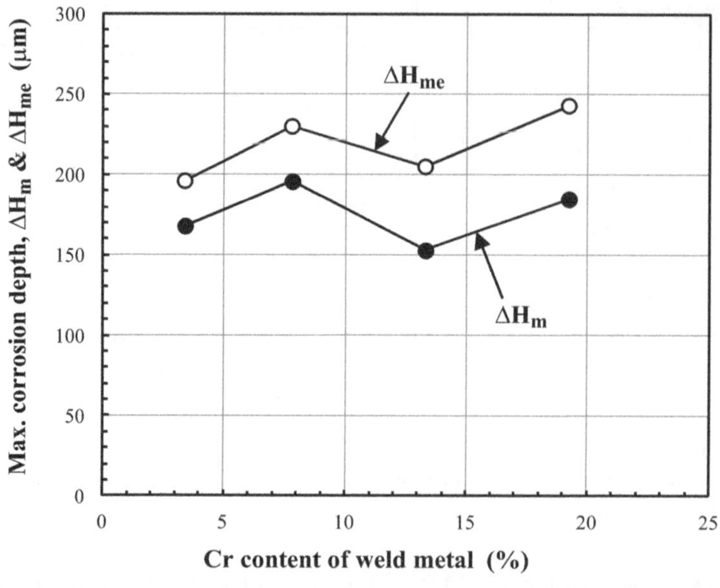

FIGURE 2.1.23
Effect of Cr content on maximum corrosion depth ΔH_m and estimated maximum corrosion depth ΔH_{me}.

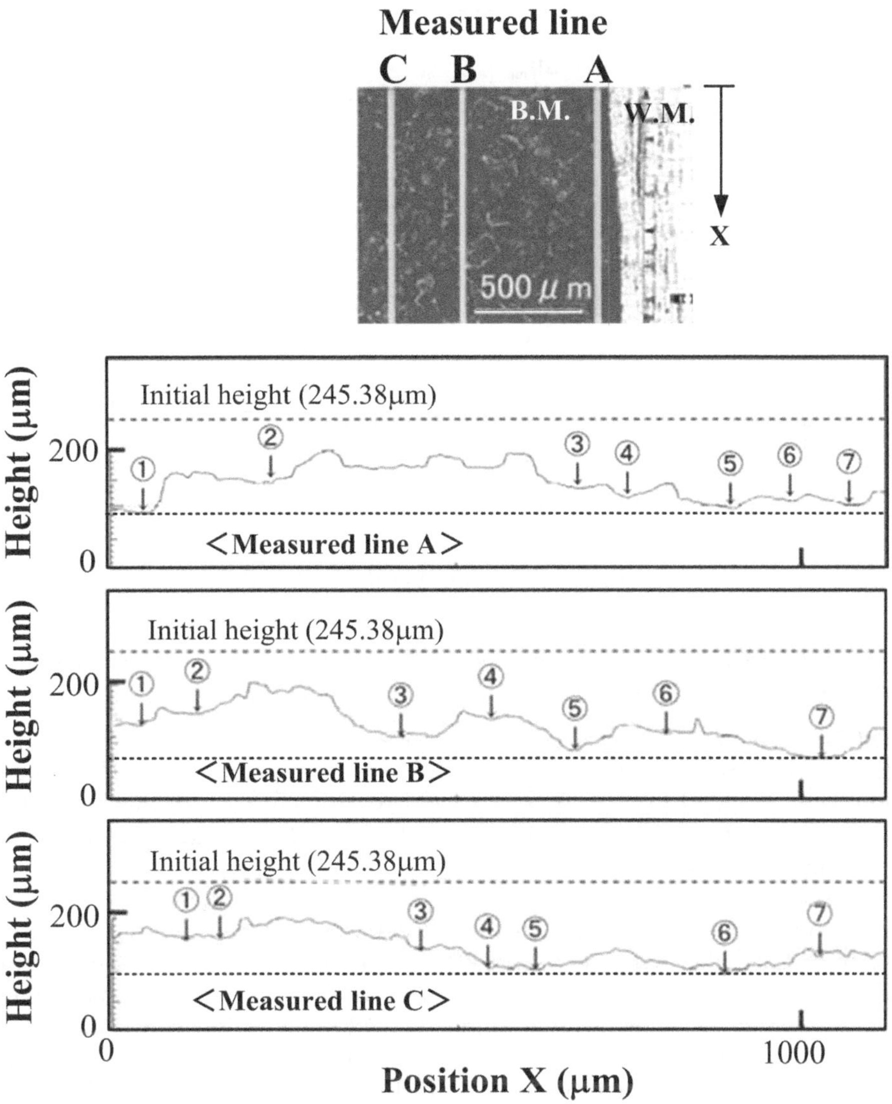

FIGURE 2.1.24
Height profile around the weld interface (No. F19).

corrosion depth, and no evidence of galvanic corrosion between the weld metal and base metal was observed.

2.1.4.2 Consideration to the Galvanic Corrosion

In atmospheric exposure tests, numerous environmental factors, such as temperature, humidity, sea salt particle deposition, and air pollutant levels,

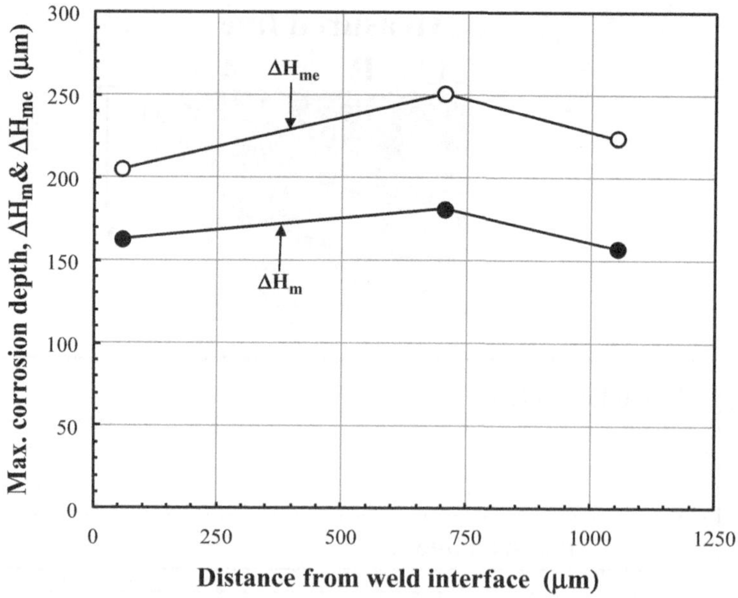

FIGURE 2.1.25
Effect of distance from the weld interface on maximum corrosion depth ΔH_m and estimated maximum corrosion depth ΔH_{me}.

fluctuate daily and complexly influence corrosion behavior. Therefore, a simplified cyclic dry/wet accelerated corrosion test was conducted. The specimens were welded joints where the weld reinforcement was mechanically removed and the weld surface was smoothed, identical to those used in the atmospheric exposure tests.

The test method involved applying a 0.5 mass% NaCl aqueous solution (0.4 L/m^2) to the specimen surface, followed by drying in a constant-temperature and -humidity chamber at 30°C and 60% RH for 12 hours. To prevent salt accumulation over time, the specimens were rinsed with pure water immediately before applying the corrosion solution. The 12-hour period constituted one cycle, and after up to 20 cycles, the rust on the specimen surface was removed by acid cleaning. The corrosion state near the fusion boundary was observed using a laser microscope, and surface height profile measurements were taken.

After subjecting welded joints with varying Cr content in the weld metal to 20 dry/wet cycles, the corrosion state of the base metal near the weld boundary was examined. Results showed that, similar to atmospheric exposure tests, general corrosion was observed on the base metal of all joints. Furthermore, as shown in Figure 2.1.26, even when the Cr content in the weld metal varied between 0.43% and 19.22%, the maximum corrosion depth (ΔH_m) near the fusion boundary remained nearly constant at 120–150 μm, with no evidence of galvanic corrosion behavior. This trend matched that of the atmospheric exposure tests, suggesting that the cyclic

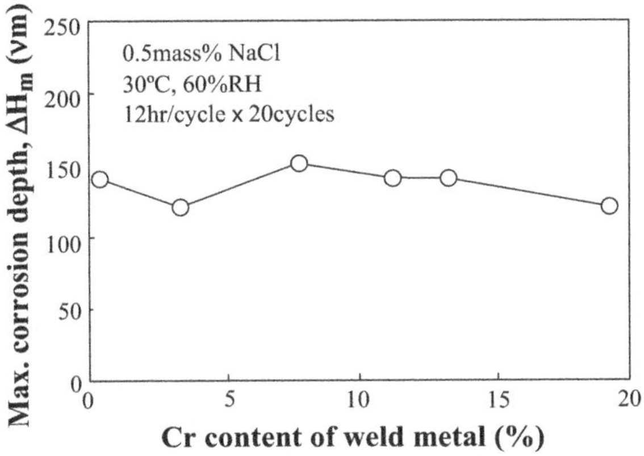

FIGURE 2.1.26
Relationship between Cr content and corrosion depth at the weld interface in a wet/dry corrosion test.

dry/wet accelerated corrosion test in NaCl solution reproduced the fundamental atmospheric corrosion mechanism.

To investigate further, the potential difference between the weld metal and base metal in NaCl solution was measured. Figure 2.1.27 shows the results for a representative case in 3mass% NaCl solution. The weld metal's potential increased proportionally with Cr content, confirming a clear potential difference with the base metal. However, since this potential difference was not reflected in the results of Figure 2.1.26, it can be concluded that the observed corrosion was not due to galvanic corrosion between the weld metal and base metal.

Regarding the atmospheric corrosion mechanism of rusted steel, it has been reported that the anodic and cathodic reactions within the rust layer

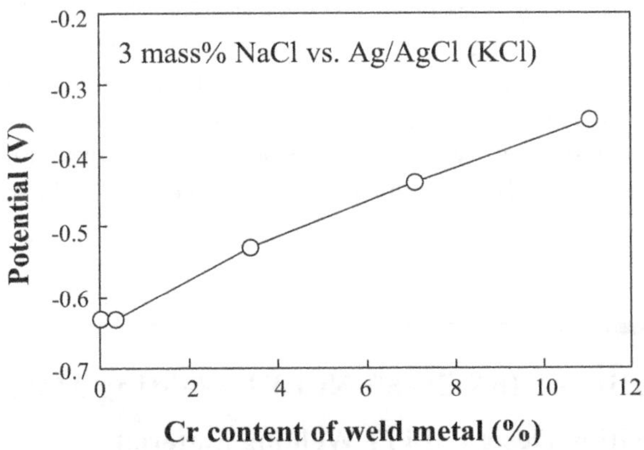

FIGURE 2.1.27
Relationship between potential and Cr content of weld metal in 3mass% NaCl.

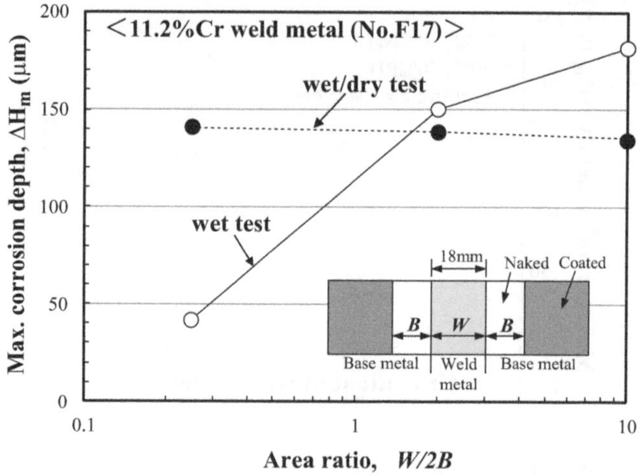

FIGURE 2.1.28
Effect of area ratio on corrosion depth in both wet and wet/dry corrosion tests.

influence corrosion behavior [36]. To examine the effect of rust formation on the surface of the base metal in relation to galvanic corrosion, an immersion corrosion test was conducted using a 3mass% NaCl solution, where stable rust layers are less likely to form due to the absence of a drying process. The results were compared with those of the cyclic dry/wet accelerated test. In the immersion test, corrosion products forming on the base metal surface were periodically removed by cleaning to actively suppress rust layer formation.

For a welded joint with 11.2%Cr–9.7%Ni weld metal (No. F17), the area ratio of weld metal to base metal was varied from 1:4 to 10:1, and a 30-day immersion corrosion test was conducted. The corrosion depth near the fusion boundary was measured and is indicated by open circles (○) in Figure 2.1.28, showing a clear difference from the results of the cyclic dry/wet test (●). As the base metal area relative to the weld metal decreased, the corrosion depth at the fusion boundary increased, exhibiting characteristics of galvanic corrosion [37–39].

Thus, while galvanic corrosion behavior was not observed in either the atmospheric exposure or cyclic dry/wet accelerated corrosion tests, the key factor is believed to be the formation of a rust layer on the base metal surface.

2.2 Weldability of 16%Cr–8%Ni LTT Welding Material

2.2.1 Composition Design of LTT Welding Material

The chemical compositions of early LTT welding material are 10%Cr–10%Ni developed by Murata et al. [9]. This composition is effective in improving fatigue strength, as demonstrated by Ota et al. [40]. However, its Cr_{eq}/Ni_{eq} ratio is approximately 1, which deviates from the recommended region shown in Figure 2.1.11.

To ensure the chemical compositions of the weld metal with dilution from the base metal fall within the recommended range, the dilution rate becomes a critical factor. The primary joints for this LTT welding material are fillet welds or overlay welds. Assuming single-layer weld metal, the dilution rate for fillet welds is estimated to be around 30%–40%, while for overlay welds, it is approximately 10%–35%.

For base metals with strengths of 490–780 MPa, the weld metal compositions must fall within the shaded gray region shown in Figure 2.2.1 at dilution rates of 10%–40%. The feasible composition range is limited to the circled area indicated by the arrow [41]. Based on these findings, the chemical compositions of the trial welding material are presented in Table 2.2.1. For reference, the compositions of the 10%Cr–10%Ni LTT welding material are also plotted in Figure 2.2.1.

Table 2.2.2 shows the calculated composition and Ms temperature of the weld metal at dilution rates of 10%–40% when using the trial 16%Cr–8%Ni welding material on 490 and 780 MPa high strength steels.

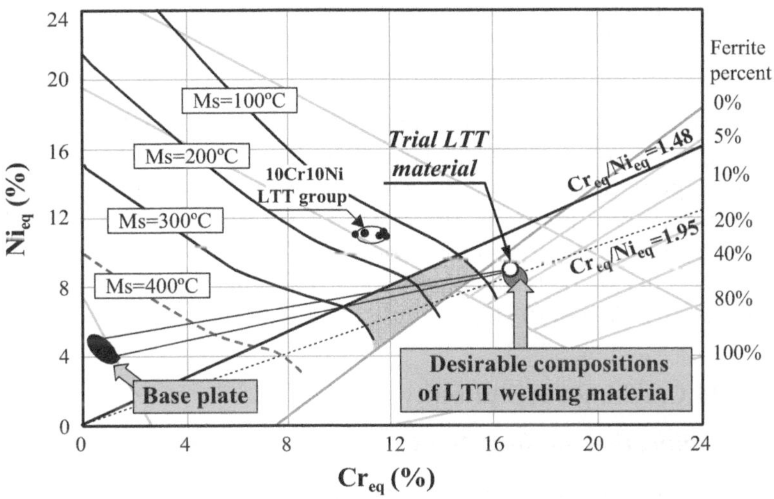

FIGURE 2.2.1
Design of chemical compositions for desirable LTT welding material.

TABLE 2.2.1
Chemical Compositions for Trial LTT Welding Material

Chemical Compositions (%)									
C	Si	Mn	P	S	Ni	Cr	Ni_{eq} (%)	Cr_{eq} (%)	Ms (°C)
0.02	0.34	0.44	0.02	0.01	8.2	16.1	8.2	16.1	29

TABLE 2.2.2

Relationship between the Dilution Rate and Weld Metal Compositions by 16%Cr–8%Ni LTT Welding Material

Dilution	Chemical Compositions (%)										
Rate (%)	C	Si	Mn	P	S	Ni	Cr	Ni_{eq} (%)	Cr_{eq} (%)	Cr_{eq}/Ni_{eq}	Ms* (°C)
10	0.03	0.33	0.51	0.02	0.01	7.4	14.5	8.6	15.0	1.74	102
20	0.05	0.33	0.57	0.02	0.01	6.6	12.9	8.2	13.4	1.63	172
30	0.06	0.32	0.64	0.02	0.01	5.7	11.3	7.8	11.8	1.50	241
40	0.07	0.32	0.70	0.02	0.01	4.9	9.7	7.4	10.1	1.36	310

* Calculated by equation "$Ms = 901 - 35.1 \cdot Cr_{eq} - 31.6 \cdot Ni_{eq}$".

(a) Base Plate: 490 MPa-Class High-Strength Steel

Dilution	Chemical Compositions (%)											
Rate (%)	C	Si	Mn	P	S	Mo	Ni	Cr	Ni_{eq} (%)	Cr_{eq} (%)	Cr_{eq}/Ni_{eq}	Ms* (°C)
10	0.03	0.32	0.48	0.02	0.01	0.05	7.5	14.5	8.6	15.0	1.75	103
20	0.04	0.30	0.52	0.02	0.01	0.10	6.8	13.0	8.1	13.4	1.65	172
30	0.04	0.28	0.56	0.02	0.01	0.15	6.1	11.4	7.7	11.8	1.51	242
40	0.05	0.26	0.60	0.01	0.01	0.20	5.4	9.9	7.3	10.3	1.41	312

* Calculated by equation "$Ms = 901 - 35.1 \cdot Cr_{eq} - 31.6 \cdot Ni_{eq}$".

(b) Base Plate: 780 MPa-Class High-Strength Steel

At a 10% dilution rate, the Ms temperature is approximately 100°C, while at 40%, it rises to around 300°C. These results confirm that the weld metal composition generally aligns with the designed recommended range.

The primary applications of the trial LTT welding material include fillet and overlay joints, as well as reinforcement and repair of existing structures. To accommodate all welding positions, it has been commercialized as an FCW (Kobelco: DW-168FP) [42].

2.2.2 Weld Cracking Susceptibility

2.2.2.1 Solidification Mode of 16%Cr–8%Ni Weld Metal

In Table 2.2.2, at a 40% dilution rate, the Cr_{eq}/Ni_{eq} ratio is less than 1.48, indicating an AF solidification mode rather than FA. This means that when using 16%Cr–8%Ni LTT with high dilution, the weld metal composition slightly deviates from the recommended range.

To experimentally confirm the primary δ phase during solidification, time-resolved X-ray diffraction using Spring-8 synchrotron radiation was performed to observe the melting and solidification process of the weld metal sample in Table 2.2.3, which deviates from the FA mode [43].

Figure 2.2.2 shows the experimental setup, where the sample was heated and melted using a TIG arc, and X-ray diffraction data was collected at

TABLE 2.2.3

Samples of Various Dilutions

Dilution Rate (%)	Chemical Compositions (%)							Ni_{eq} (%)	Cr_{eq} (%)	Cr_{eq}/Ni_{eq}	Ms[a] (°C)
	C	Si	Mn	P	S	Ni	Cr				
0	0.02	0.34	0.44	0.02	0.01	8.2	16.1	9.0	16.6	1.84	33
34	0.07	0.31	0.64	0.02	0.01	5.4	10.6	7.8	11.1	1.43	266
39	0.07	0.30	0.67	0.02	0.01	5.0	9.8	7.6	10.3	1.36	306

[a] Calculated by equation "$Ms=901-35.1 \cdot Cr_{eq}-31.6 \cdot Ni_{eq}$".

FIGURE 2.2.2
Synchrotron radiation setup for in situ observation and analysis of phase transformation.

0.02-second intervals during melting, solidification, and cooling to room temperature [43–45].

Figure 2.2.3 displays the diffraction profile at the onset of solidification. The x marks represent austenite diffraction peaks, while the ○ marks indicate δ-ferrite peaks. Although the δ-ferrite peak intensity decreases with increasing dilution, primary δ-ferrite is detected up to a 39% dilution rate, confirming an FA solidification mode. Thus, FA mode solidification is maintained at dilution rates of 0%–40%.

2.2.2.2 Verification of Weld Hot Cracking Susceptibility in the FISCO Test

One method for assessing hot cracking susceptibility is the "C-restraint weld cracking test (FISCO test)". This test was conducted using the 16%Cr–8%Ni LTT welding material (Kobelco: DW-168FP), with reference data from a 20%Cr–10%Ni commercial stainless steel welding material (Kobelco: DW-308L). Both materials are FCWs. For the FISCO test, the butt gap was set to 2.6 mm, with welding conditions of 200 A and 28 V.

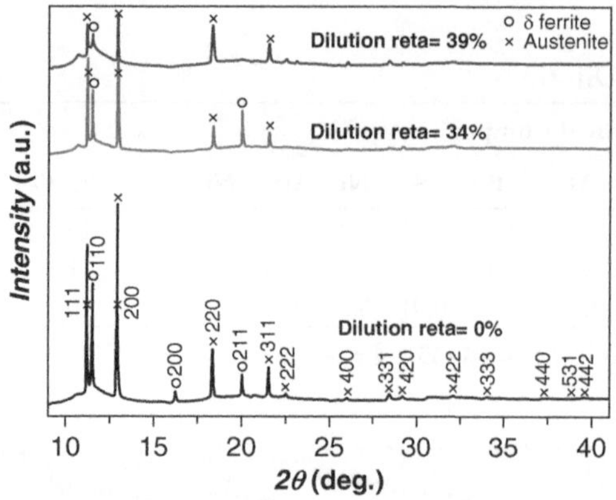

FIGURE 2.2.3
In situ X-ray diffraction pattern of the initial phase transformation during solidification for samples of various dilutions [43].

The cracking length in the bead and crater regions (Figure 2.2.4) is summarized in Figures 2.2.5 and 2.2.6, respectively. The cracking ratio is defined as

- Crater region cracking ratio (%) = (Crater crack length/L_{crater}) × 100
- Bead region cracking ratio (%) = (Bead crack length/L_{bead}) × 100
- Total cracking ratio (%) = (Total crack length/L_{total}) × 100

Figure 2.2.5 shows no hot cracking in the bead region for either the 16% Cr–8%Ni or 20%Cr–10%Ni weld metal, even with varying welding speeds. However, Figure 2.2.6 reveals a high cracking ratio in the crater region for both, though their relative resistance to hot cracking is unclear. The overall cracking rate (bead + crater) is shown in Figure 2.2.7 [42].

These results indicate no significant difference in hot cracking resistance between the two weld metals, confirming that the 16%Cr–8%Ni LTT weld metal performs comparably to the commercial 20%Cr–10%Ni weld metal.

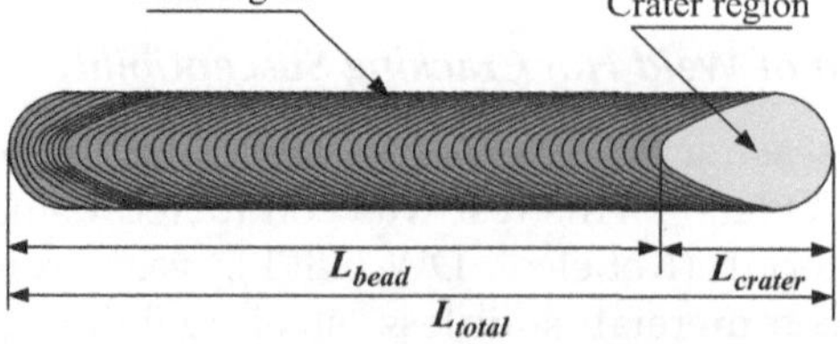

FIGURE 2.2.4
Definition of the welding bead region.

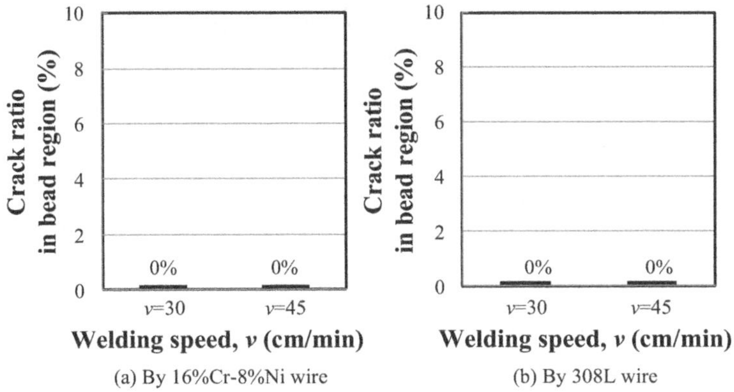

FIGURE 2.2.5
Crack ratio in the bead region.

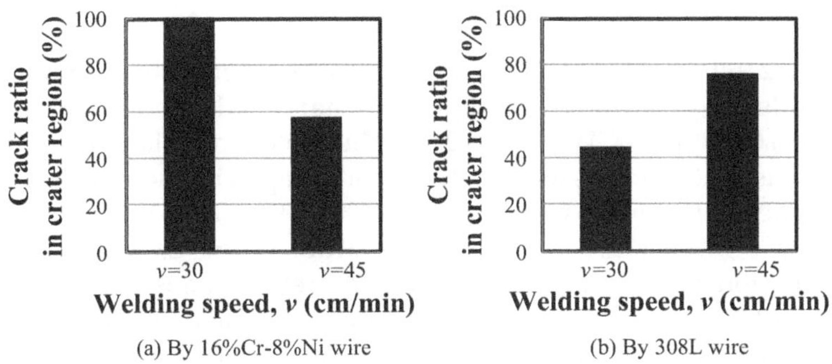

FIGURE 2.2.6
Crack ratio in the crater region.

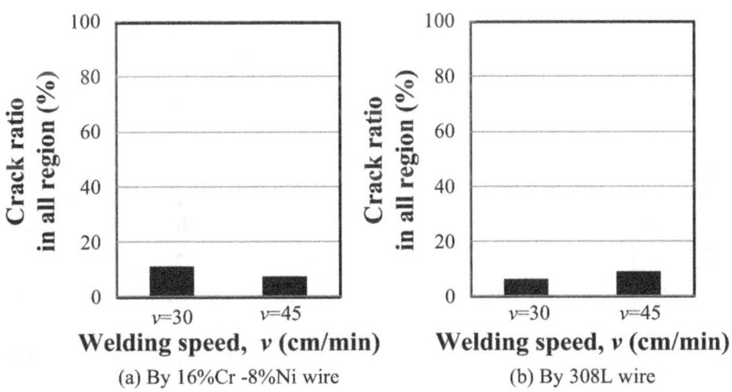

FIGURE 2.2.7
Crack ratio in the FISCO test.

2.2.2.3 Examination of Weld Cold Cracking Susceptibility of 16%Cr–8%Ni Weld Metal

As noted in Section 2.1.2, reducing cold cracking susceptibility in LTT weld metal involves:

1. Reducing tensile residual stress or generating compressive residual stress due to apparent martensitic transformation expansion.
2. Utilizing retained austenite to trap hydrogen and reduce diffusible hydrogen content, a weld metal with a dual-phase structure (martensite+retained austenite) is recommended.

Figure 2.2.8 presents the results of the y-groove weld cracking test (Tekken test) using 10%Cr–10%Ni FCW wire on HT780 high strength steel [46]. The measured Ms temperature was 269°C, with a 0% cracking ratio, demonstrating reduced cold cracking susceptibility with LTT FCW. This data was used to compare the cold cracking susceptibility of 16%Cr–8%Ni FCW with that of 10%Cr–10%Ni FCW.

Figure 2.2.9 shows the relationship between apparent transformation strain ($\Delta\varepsilon$) and Ms temperature. In this figure, ○ marks indicate the measured values for 16%Cr–8%Ni FCW. Figure 2.2.10 illustrates the relationship between retained austenite and Ms temperature. Both graphs include the dilution rate for weld metals using 16%Cr–8%Ni and 11%Cr–9%Ni FCW on HT780 steel.

For 11%Cr–9%Ni FCW, the Ms temperature is 269°C, with an apparent transformation strain of 0.43% and about 2% retained austenite at a dilution rate of about 37%. Under the same conditions, weld metal using 16%Cr–8%Ni FCW has an Ms temperature of 280°C, a similar transformation strain (0.43%), and retained austenite (1.5%). If the hydrogen content is equal for both LTT welding wires, the y-groove test should also yield a 0% cracking ratio at a dilution rate of≤37%. Thus, it can be concluded that 16%Cr–8%Ni FCW has sufficient resistance to cold cracking.

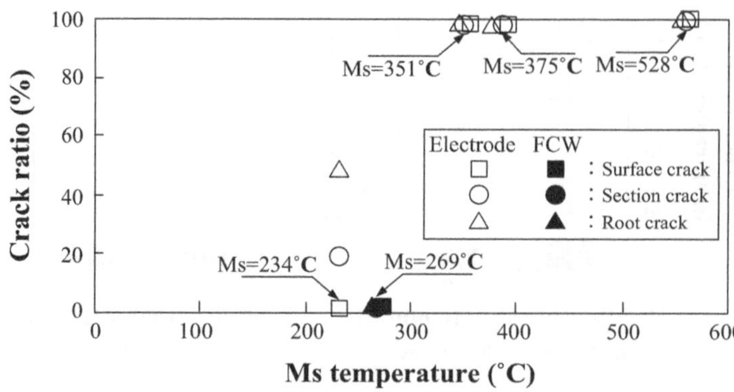

FIGURE 2.2.8
Relationship between the crack ratio and Ms temperature in the y-groove cracking test.

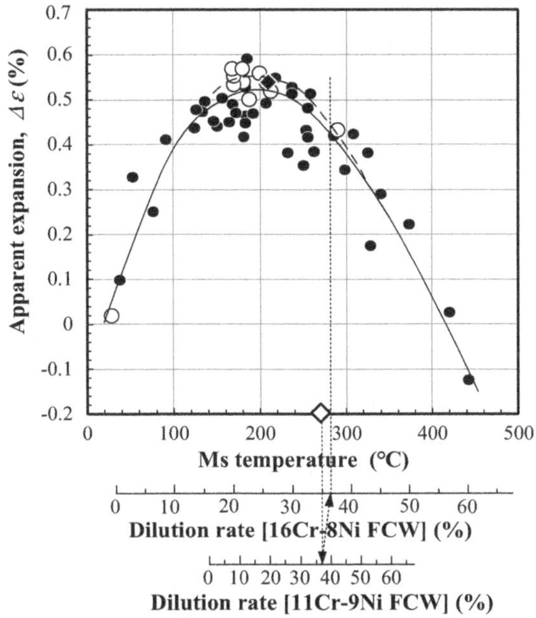

FIGURE 2.2.9
Relationship among apparent expansion, Ms temperature, and dilution rate (Base plate: 780 MPa-class high strength steel).

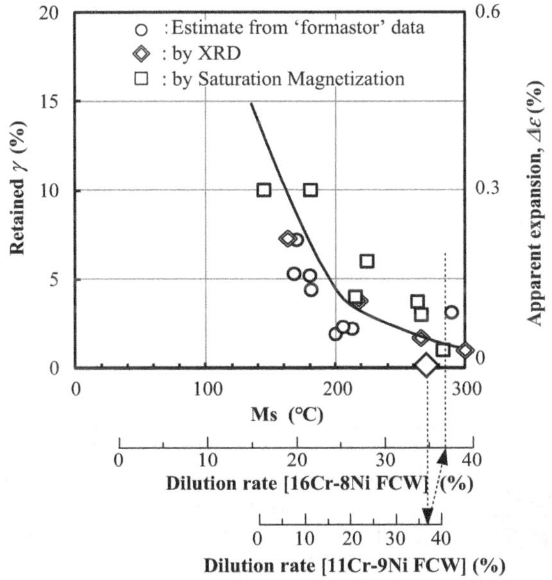

FIGURE 2.2.10
Relationship among retained γ, Ms temperature, and dilution rate (Base plate: 780 MPa-class high strength steel).

2.2.2.4 *Verification of Weld Cold Cracking Susceptibility in Restrained Fillet Crack Test*

Cold cracking resistance was evaluated using a restrained fillet weld cracking test [47], comparing 16%Cr–8%Ni LTT FCW with a commercial LTT FCW for bridges.

Figure 2.2.11 shows the test specimen (490 MPa steel plate), assembled following standard procedures using conventional welding wire.

Figure 2.2.12 confirms that no cold cracks were found in the fillet weld metal after 48 hours, as determined by magnetic particle testing.

The measured Ms temperature was 290°C for 16%Cr–8%Ni LTT weld metal with a 37% dilution rate, 0.43% apparent transformation strain, and 1%–2% retained austenite, which are similar to the 10%Cr–10%Ni FCW results, validating low cracking susceptibility.

Figure 2.2.13 compares additional bead welding using 16%Cr–8%Ni FCW and commercial bridge LTT FCW [48]. No cracks were observed in

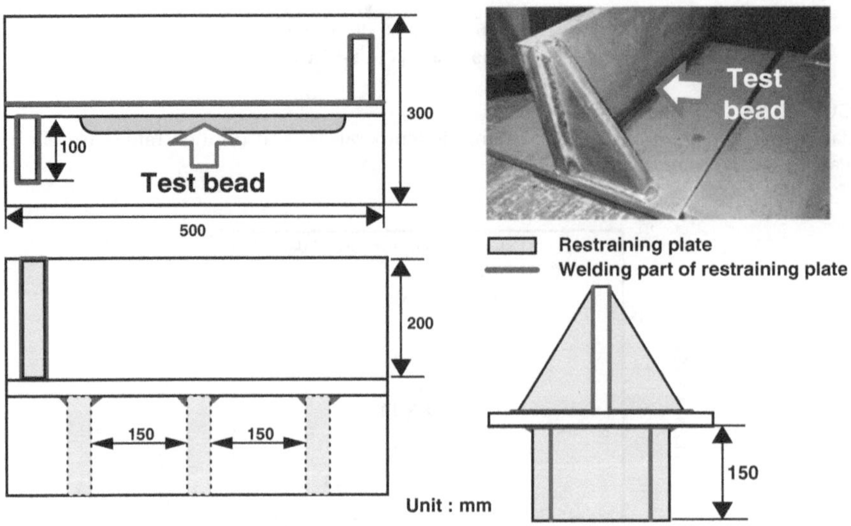

FIGURE 2.2.11
Shapes of the test assembly.

Joint Style	LTT Wire	Part	Initial temperature (accuracy: ±5°C)				
			0°C	25°C	50°C	75°C	100°C
	Developed 16Cr-8Ni	Surface of bead	No	No	No		
		Inside of bead	No	No	No		

FIGURE 2.2.12
Results of the cold cracking test proposed by Class NK with 16%Cr–8%Ni FCW.

Joint Style	LTT Wire	Part	Initial temperature (accuracy: ±5°C)				
			0°C	25°C	50°C	75°C	100°C
	Developed 16Cr-8Ni	Surface of bead	No	No	No		
		Inside of bead	No	No	No		
	Conv. LTT	Surface of bead	**Crack**	**Crack**	No	No	No
		Inside of bead	–	–	No	No	No

FIGURE 2.2.13
Results of the cold cracking test proposed by Class NK with 16%Cr–8%Ni FCW or LTT FCW for bridges.

16%Cr–8%Ni welds at a 37% dilution rate without preheating, whereas the commercial FCW required ≥50°C preheating to prevent cracking [42].

In conclusion, the 16%Cr–8%Ni LTT FCW exhibits excellent cold cracking resistance, attributed to hydrogen trapping by retained austenite and compressive residual stress from the expansion of martensitic transformation.

2.2.3 Tensile Properties and Hardness

2.2.3.1 Tensile Properties of 16%Cr–8%Ni LTT Weld Metal

2.2.3.1.1 Tensile Properties of the as-Welded Miniature Specimen

Figure 2.2.14 illustrates the experimental setup and specimen dimensions for the miniature uniaxial tensile test. The specimens were extracted from the center of 16%Cr–8%Ni LTT welds with their length parallel to the welding direction. The dilution rates of the 16%Cr–8%Ni LTT welds were 27%, 34%, and 39%, respectively. The uniaxial tensile tests were performed on miniaturized specimens with dimensions of 1.0 mm (length), 0.4 mm (width), and 0.2 mm (thickness), using a crosshead speed of 2 μm/s.

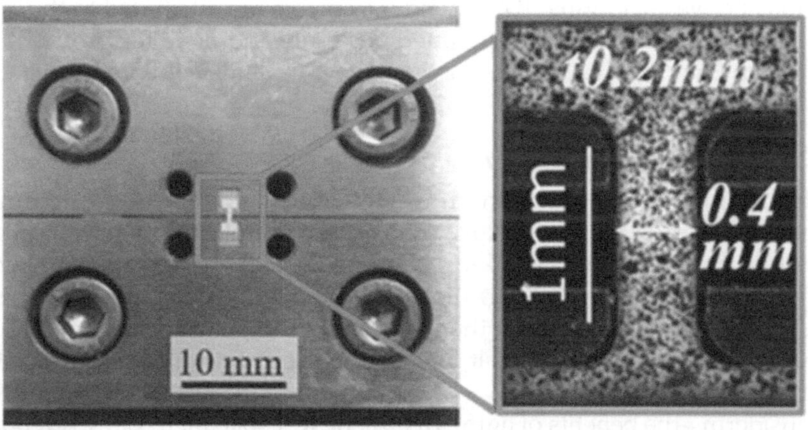

FIGURE 2.2.14
Experimental setup for the miniature uniaxial tensile test.

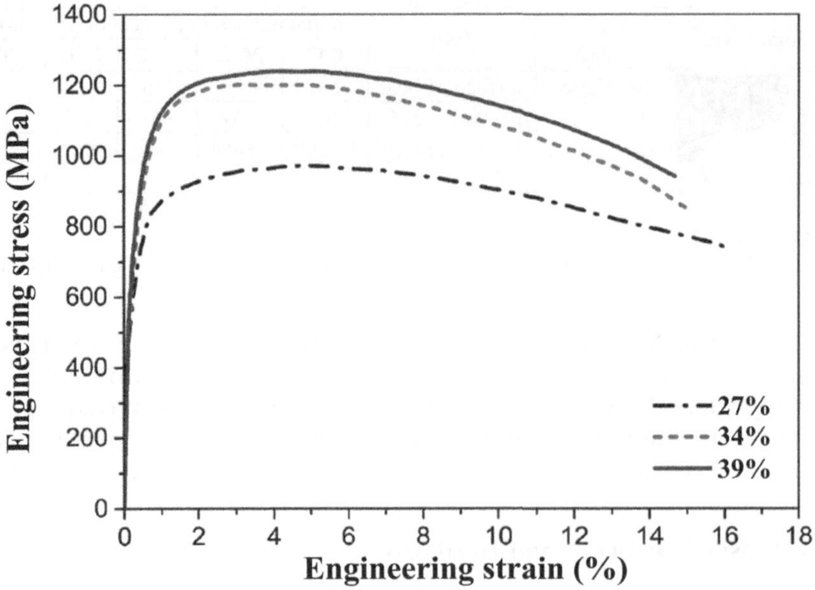

FIGURE 2.2.15
Engineering stress–strain curves of different LTT weld metals.

Figure 2.2.15 shows the stress–strain curves of 16%Cr–8%Ni LTT weld metals with dilution rates of 27%, 34%, and 39%, respectively. The 0.2% offset yield strengths (YS /0.2%PS) of these three 16%Cr–8%Ni LTT weld metals were 812, 955, and 1,070 MPa, respectively, while the tensile strength (TS) values were 973, 1,202, and 1,242 MPa, respectively. Additionally, the elongation (EL) values for these weld metals were 16%, 15%, and 14.6%, respectively. As dilution increased, both the YS and TS increased, while the variation in EL of about 1% was negligible. This trend was primarily attributed to the reduction in the volume fraction of retained austenite and delta ferrite as dilution increased. Overall, the 16%Cr–8%Ni LTT weld metals exhibited satisfactory tensile properties.

2.2.3.1.2 Tensile Properties of Standard Specimen

In the production of high strength welded joints, manufacturers commonly employ aging treatment to reduce hydrogen concentration and mitigate the risk of hydrogen-induced cracking. Moreover, as demonstrated by Wu et al. [49,50], aging treatment can significantly enhance the mechanical properties of LTT weld metals. On the other hand, low-temperature aging treatment (<150°C) is often preferred, as it has minimal impact on the residual stress distribution.

Considering the benefits of aging treatment, test specimens of LTT deposited metal and weld metal with a dilution rate of 27%, as shown in Figure 2.1.16a and c, were subjected to aging treatment at 110°C for 12 hours. Additionally, an as-welded LTT weld metal with a dilution rate of 20%, as shown in Figure 2.1.16b, was prepared to investigate the effects of aging treatment.

The engineering stress–strain curves of 16%Cr–8%Ni LTT deposited metal and weld metals are shown in Figure 2.2.17 and summarized in Table 2.2.4. According to the relationship between dilution rate and retained austenite content, 16%Cr–8%Ni LTT weld metal with a lower dilution rate typically exhibits a higher percentage of retained austenite, which is generally beneficial for EL. However, the EL of the 16%Cr–8%Ni LTT weld metal with a 20% dilution rate is lower than that of the weld metal with a 27% dilution rate. This observation indicates that the round bar shape (as opposed to plate) of

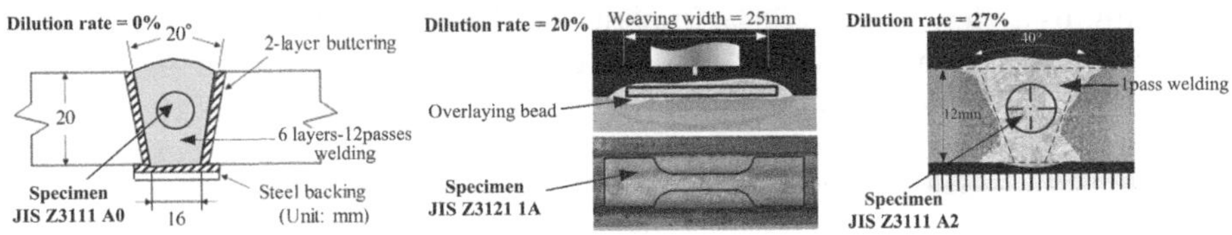

(a) Deposit metal (b) Weld metal in overlaying (c) Weld metal in butt joint

FIGURE 2.2.16
Groove shape and sampling of specimens.

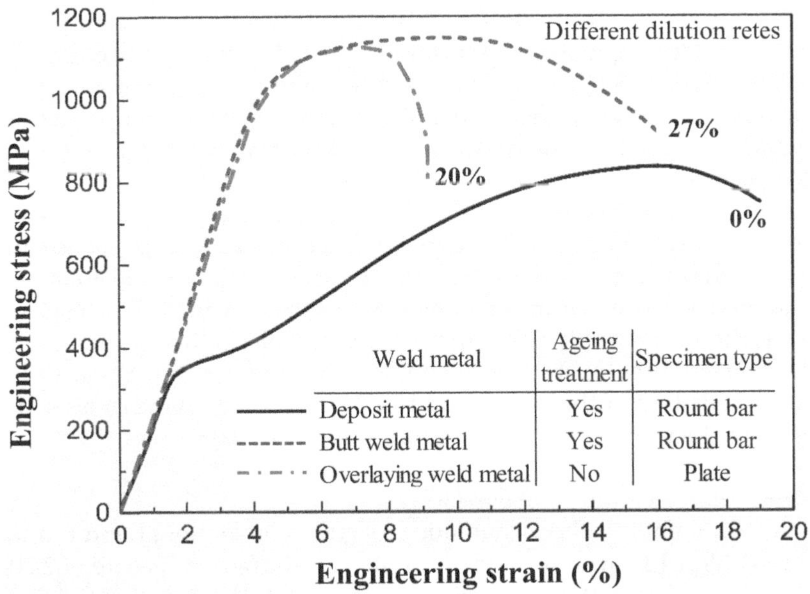

FIGURE 2.2.17
Engineering stress–strain curves of 16%Cr–8%Ni LTT deposited metal and weld metals (without using extensometers).

TABLE 2.2.4

Tensile Properties of 16%Cr–8%Ni LTT Deposited Metal and Weld Metals

Dilution Rate (%)	YS/0.2%PS (MPa)	TS (MPa)	EL (%)	Aging Treatment
0	845	845	19	Yes
20	1,123	1,123	9	No
27	890	1,148	16	Yes

the tensile test specimen and low-temperature aging treatment can significantly enhance the EL of LTT weld metal.

Meanwhile, the 16%Cr–8%Ni LTT deposited metal demonstrates excellent strength and ductility, with a YS of 321 MPa, TS of 845 MPa, and EL of 19%, resulting in a yield ratio of 38%. In contrast, the weld metal with a 27% dilution rate exhibits a YS of 890 MPa, TS of 1,148 MPa, and EL of 16%, corresponding to a yield ratio of 76%. These results highlight the high strength and ductility of the weld metal with a 27% dilution rate. It is evident that both YS and TS increase with the dilution rate.

2.2.3.2 Ductility of 16%Cr–8%Ni LTT Welded Joints

2.2.3.2.1 Ductility of Butt Joint

The aforementioned findings highlight that the as-welded LTT weld metal exhibits relatively poor EL. Despite this, the LTT welding wire can still be considered suitable for practical applications, provided that the strength–ductility balance (tensile toughness) of the welded joint is equivalent to or superior to that of conventional welded joints. To this end, the tensile properties of both the 16%Cr–8%Ni LTT butt-welded joint and overlay-welded joint have been investigated.

In accordance with the standard JIS Z 2241, a static tensile test was conducted on a butt joint, with the tensile direction oriented perpendicular to the weld length. Figure 2.2.18 presents the test results for the 16%Cr–8%Ni LTT welded joint, revealing YS and TS of 409 and 546 MPa, respectively. Notably, the EL reached 25%, which is comparable to that of the base metal KA36 steel. As previously mentioned, the TS of the 16%Cr–8%Ni LTT weld metal exceeds that of the base metal KA36, indicating an overmatched joint. Ultimately, ductile fracture occurred in the base metal.

2.2.3.2.2 Ductility of LTT Overlay Joint

Figure 2.2.19 shows the schematics of the conventional joint and the 16%Cr–8%Ni LTT overlay-welded joint. As shown in Figure 2.2.20, the test results indicate that the TS and ELs of the conventional joint and the 16%Cr–8%Ni LTT overlay-welded joint are 510 MPa (31%) and 610 MPa (21%), respectively [51]. Although the EL of the 16%Cr–8%Ni LTT overlay-welded joint is lower than that of the conventional joint, it still meets the application requirements for ship hull structures.

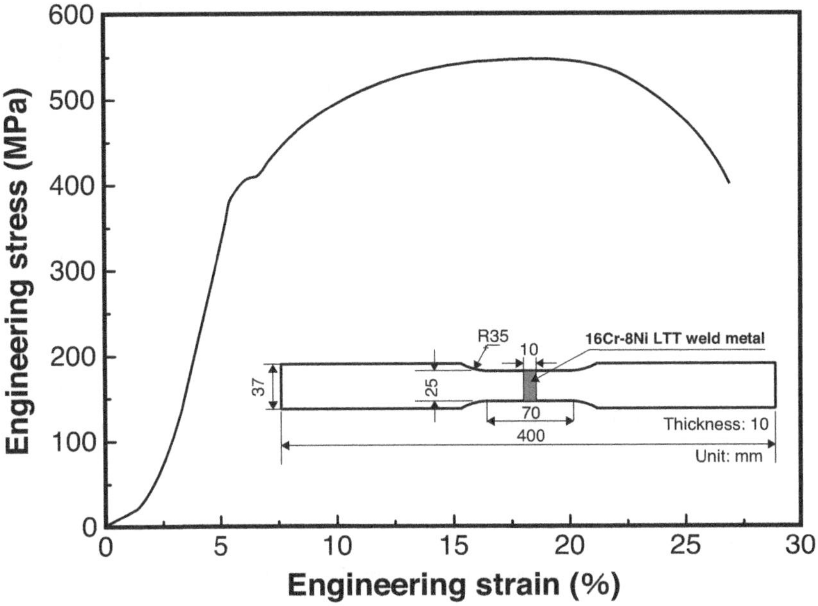

FIGURE 2.2.18
Engineering stress–strain curve of the 16%Cr–8%Ni LTT welded joint.

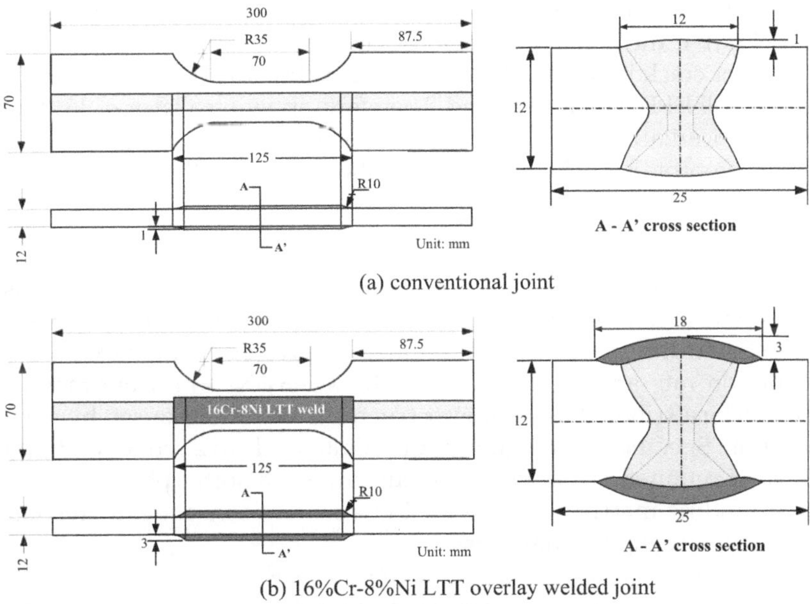

FIGURE 2.2.19
Schematic of welded joints.

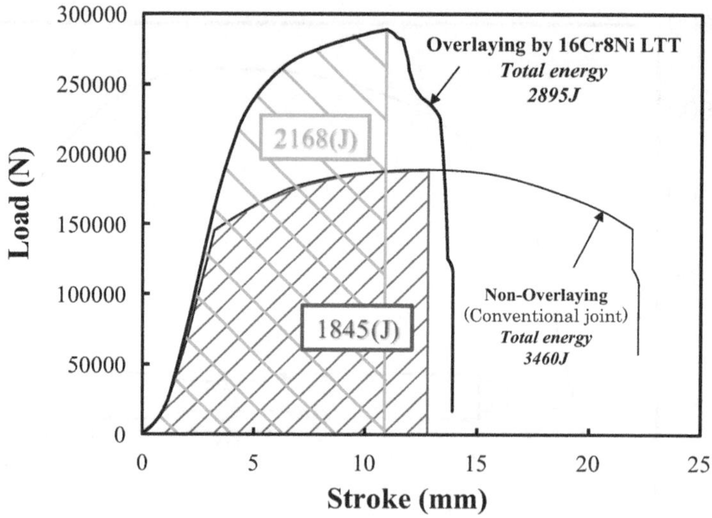

FIGURE 2.2.20
Tensile curves of the conventional joint and LTT overlay-welded joint.

In terms of deformation energy, the 16%Cr–8%Ni LTT overlay specimen exhibits a higher deformation energy up to the maximum load (2.17 kJ) compared to the conventional butt specimen made with conventional welding wire (Kobelco: MXZ-200) (1.85 kJ). However, the total deformation energy of the 16%Cr–8%Ni LTT overlay joint is 2.90 kJ, which is slightly lower than that of the conventional joint (3.46 kJ). If the deformation energy up to the maximum load is used as a criterion for crack initiation, the risk of ductile fracture is relatively low for the 16%Cr–8%Ni LTT overlay cladding. Therefore, applying a layer of 16%Cr–8%Ni LTT metal overlay cladding can generate beneficial compressive residual stresses without compromising the strength–ductility balance of the conventional joint.

2.2.3.3 Hardness

The hardness of the 16%Cr–8%Ni LTT weld metals was measured using a Vickers hardness tester, with a load of 2 kg and a dwell time of 15 seconds. As illustrated in Figure 2.2.21, a distinct positive correlation exists between the dilution rate and the hardness of 16%Cr–8%Ni LTT weld metals. This trend can be attributed to the microstructural evolution caused by variations in dilution. Specifically, a higher dilution rate leads to a reduction in the percentage of retained austenite. Since austenite is a softer phase compared to its surrounding matrix, the decrease in its content consequently results in an overall increase in weld hardness.

2.2.4 Fracture Toughness of LTT Weld Metal

LTT weld metal primarily consists of a martensitic microstructure, which may have an issue concerning low toughness. However, as mentioned in

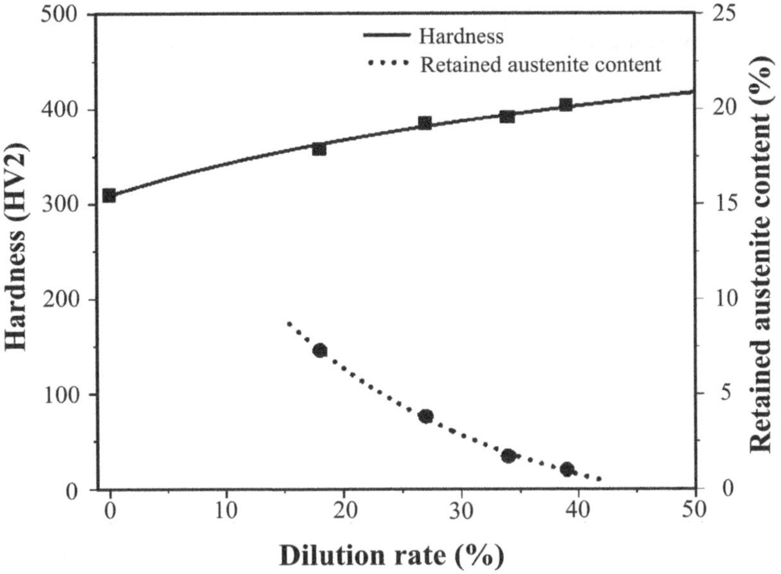

FIGURE 2.2.21
Effect of dilution rate on hardness and retained austenite content.

Section 2.1.3, even with a predominantly martensitic structure, it is possible to improve the toughness of the weld metal by employing welding techniques that suppress oxide formation.

Nevertheless, in practical applications, there are various constraints, including requirements and demands that depend on the working environment (e.g., in-factory or on-site construction).

The 16%Cr–8%Ni LTT welding material was developed with the premise of being applied to repair cracks in welded joints of ship structures. Therefore, it is essential to meet construction requirements for all welding positions and weldability with 100% CO_2 shielding gas for lower costs. For these purposes, an FCW was commercialized as DW-168FP by Kobelco.

Before the 16%Cr–8%Ni LTT welding material, another 10%Cr–10%Ni LTT welding material as the solid wire was also manufactured and used with 80%Ar–20%CO_2 shielding gas in MAG welding, typically in the flat welding position.

This section evaluates the toughness of 16%Cr–8%Ni LTT weld metal in comparison with 10%Cr–10%Ni LTT weld metal, based on the results of V-notch Charpy impact tests and center-notched wide plate tensile tests.

2.2.4.1 Fabrication of Butt-welded Joints

The base metal of the butt-welded joint is a 490 MPa-class high strength steel plate (KA36: thickness 12mm) whose chemical composition and mechanical properties are shown in Table 2.2.5. Figure 2.2.22 illustrates the groove shape and penetration of the butt-welded joint, and Table 2.2.6 presents the corresponding welding conditions. For comparison, joints were also fabricated using general carbon steel welding wire (Kobelco: MX-Z200) for high strength steel.

TABLE 2.2.5

Chemical Composition and Mechanical Properties of Steel Plate

	Chemical Compositions					Mechanical Properties			
	C	Si	Mn	P	S	YS	TS	EL	$_vE_{0°C}$
Steel	(%)	(%)	(%)	(%)	(%)	(N/mm²)	(N/mm²)	(%)	(J)
KA36 (12 mmt)	0.16	0.24	1.04	0.015	0.005	418	525	24	209

TABLE 2.2.6

Welding Parameters of Test Specimens

Wire	Welding Method	Shielding Gas	Welding Run	Welding Current (A)	Arc Voltage (V)	Welding Speed (cm/min)	Heat Input (kJ/cm)
Conventional (FCW)	GMAW	100%CO_2	1st	150	29	30	8.7
			2nd	300	30	25	21.6
10Cr10Ni LTT (Solid wire)		80%Ar 20%CO_2	1st	150	24	30	7.2
			2nd	330	34	25	26.9
16Cr8Ni LTT (FCW)		100%CO_2	One-run	150	24	6	36

Gas flow rate: 25 L/min, interpass temperature: <150°C.

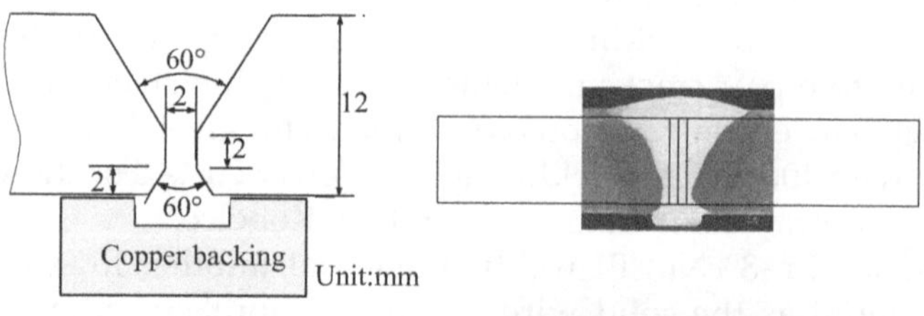

(a) In case of 10%Cr-10%Ni LTT solid wire and conventional FCW.

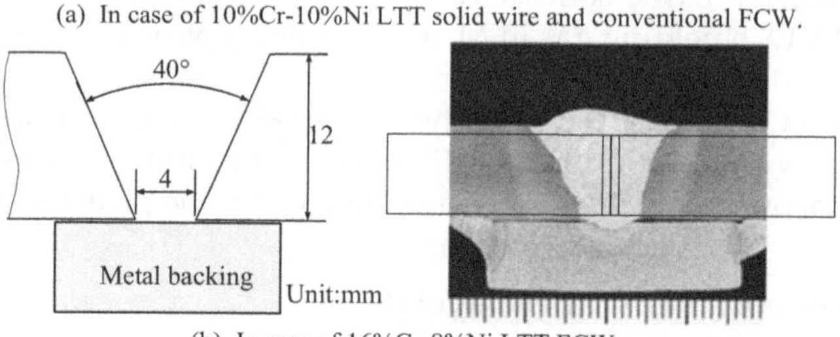

(b) In case of 16%Cr-8%Ni LTT FCW.

FIGURE 2.2.22
Groove shape and penetration shape in butt-welded joints.

TABLE 2.2.7

Dilution Rate, Chemical Composition, and Properties of LTT Weld Metals

Weld Metal	Dilution rate (%)	Chemical Composition (%)										Solid. Mode	Ms (°C)
		C	Si	Mn	P	S	Ni	Cr	Ni_{eq}	Cr_{eq}	Cr_{eq}/Ni_{eq}		
10Cr10Ni LTT	22	0.07	0.26	0.67	0.02	0.01	7.4	8.2	9.9	8.6	0.87	A	221
16Cr8Ni LTT	27	0.06	0.31	0.60	0.02	0.01	6.0	11.8	8.0	12.2	1.52	FA	218

For the groove shown in Figure 2.2.22a, a two-pass welding technique was employed, with one pass on the back side and one pass on the front side. However, when welding with the 10%Cr–10%Ni solid wire, the front-side welding caused remelting of the first pass on the back side, effectively making it equivalent to single-run welding.

The dilution rate, chemical compositions, and other properties of the two LTT weld metals are summarized in Table 2.2.7.

2.2.4.2 V-Notch Charpy Impact Test Results of LTT Weld Metal

For the 10%Cr–10N%i LTT weld metal cut from the center cross-section of the butt-welded joint, the absorbed energy and temperature transition curve from standard V-notch Charpy impact tests are presented in Figure 2.2.23 [52]. Then, scanning electron microscopy (SEM) images of the fracture surfaces at each test temperature are shown in Figure 2.2.24. Even at –180°C and below –120°C, the fracture surface exhibits not only brittle fracture but also over 50% dimple fracture.

The V-notch Charpy impact test results for the 10%Cr–10%Ni LTT weld metal with a 22% dilution rate show that the 50% absorbed energy transition temperature (vTrE) is significantly low (around –120°C), although the upper shelf energy, approximately 50 J, is also low.

For the 16%Cr–8%Ni LTT weld metal cut from the center cross-section of the butt-welded joint, the absorbed energy and temperature transition curve from standard V-notch Charpy impact tests are presented in Figure 2.2.25 [53]. As illustrated in Figure 2.2.25, the test specimens of the 16%Cr–8%Ni LTT weld metal included both as-welded specimens and those subjected to aging heat treatment. The aging heat treatment condition was set at 100°C gas-shielded for 24 hours. It is clear from Figure 2.2.25 that the weld metals in both the as-welded and aged states exhibited the same absorbed energy characteristics.

Due to the use of 100% CO_2 gas-shielded arc welding, the absorbed energy at 0°C ($_{V}E_{0°C}$) for the 16%Cr–8%Ni LTT weld metal is 31 J, which is lower compared to the 10%Cr–10%Ni LTT weld metal. Additionally, the 50% absorbed energy transition temperature (vTrE) is approximately –120°C for the 10%Cr–10%Ni LTT weld metal, whereas it is around –100°C for the 16%Cr–8%Ni LTT weld

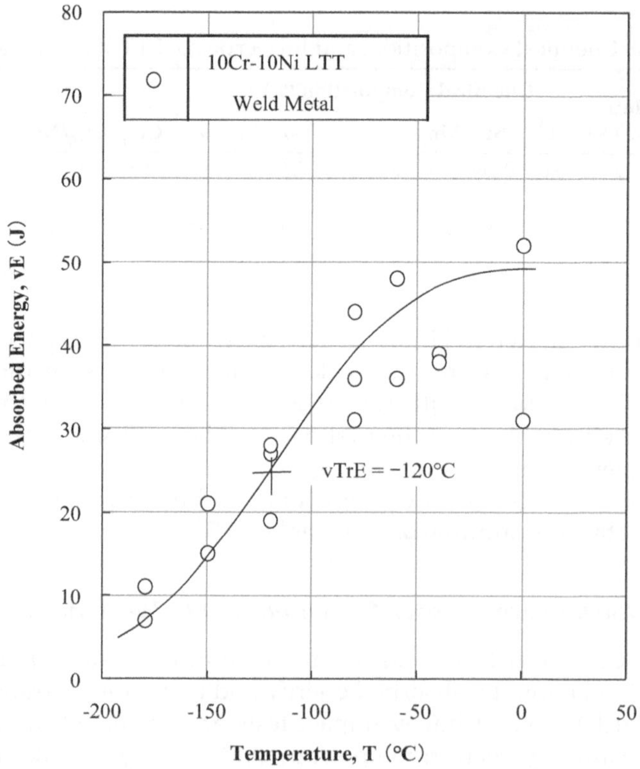

FIGURE 2.2.23
Results of V-notch Charpy impact tests in 10%Cr–10%Ni LTT weld metal.

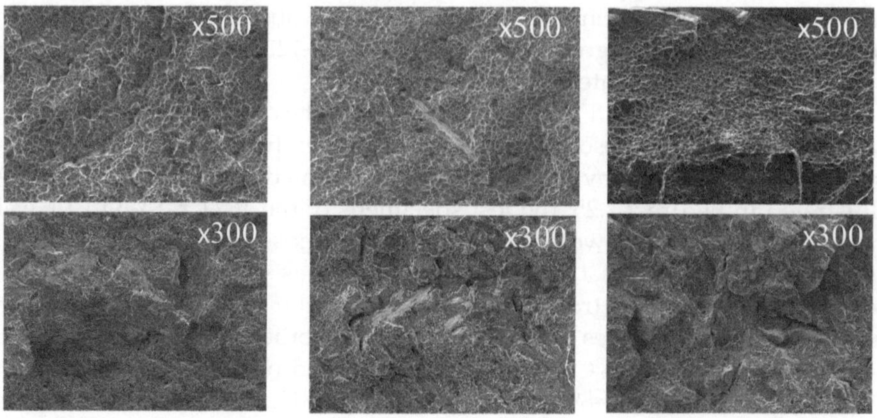

(a) Test temperature −120°C (Percent brittle fracture:17.5%)
(b) Test temperature −150°C (Percent brittle fracture:21%)
(c) Test temperature −180°C (Percent brittle fracture:33%)

FIGURE 2.2.24
Fracture surface after Charpy impact tests in 10%Cr–10%Ni LTT weld metal.

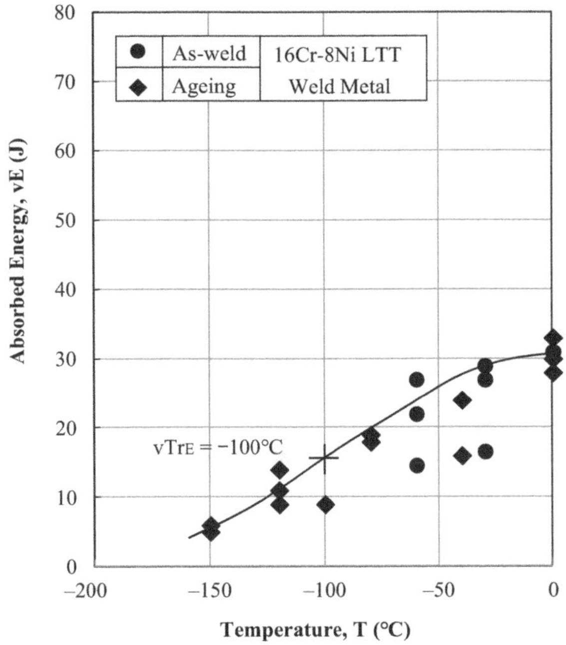

FIGURE 2.2.25
Results of V-notch Charpy impact tests in 16%Cr–8%Ni LTT weld metal.

metal. This suggests that the fracture toughness of the 10%Cr–10%Ni LTT weld metal is slightly better than that of the 16%Cr–8%Ni LTT weld metal.

Figure 2.2.26 shows SEM images of fracture surfaces at various test temperatures. Up to –120°C, the fracture surface primarily exhibits dimple patterns, but at around –150°C, the dimple fractures clearly decrease, and brittle fractures dominate. Observations of the fracture surfaces at low temperatures suggest that the 10%Cr–10%Ni LTT weld metal deposited under an 80%Ar–20%CO_2 shielding gas exhibits slightly superior fracture toughness.

However, when welding is performed using a 16%Cr–8%Ni LTT wire with an 80%Ar–20%CO_2 shielding gas, as described in Section 2.1.3, the upper shelf energy and the 50% absorbed energy transition temperature (vTrE) are presumed to be nearly equivalent to those of welds made with the 10%Cr–10%Ni LTT wire.

Table 2.2.8 shows the 50% absorbed energy transition temperature (vTrE) from V-notch Charpy impact tests for the fusion zones (i.e. region between weld and HAZ) of electroslag welds and submerged arc welds in 490 MPa grade shipbuilding steel plates, which represent the most brittle regions [54]. Although the V-notch Charpy absorbed energy of LTT weld metals, as shown in Figures 2.2.23 and 2.2.25, is significantly lower than that of the fusion zones in conventional carbon steel electroslag welds and submerged arc welds, the vTrE of LTT weld metals is extremely low, below –100°C. In this regard, the fracture toughness of LTT weld metals can be considered superior to that of conventional carbon steel weld metals.

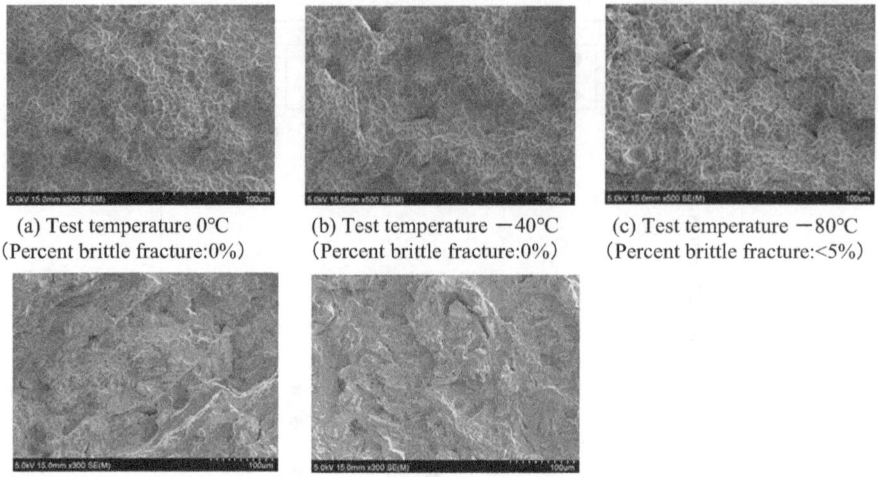

(a) Test temperature 0°C (Percent brittle fracture:0%)
(b) Test temperature −40°C (Percent brittle fracture:0%)
(c) Test temperature −80°C (Percent brittle fracture:<5%)
(d) Test temperature −120°C (Percent brittle fracture:36%)
(e) Test temperature −150°C (Percent brittle fracture:68%)

FIGURE 2.2.26
Fracture surface after Charpy impact tests in 16%Cr–8%Ni LTT weld metal.

TABLE 2.2.8

Results of Transition Temperature in V-Notch Charpy Impact Tests at LTT Weld Metal and at the Fusion Line of Electroslag Weld Joint and Submerged Arc Weld Joint by Using Conventional Steel Wire

Evaluated Part	Welding Wire	Joint	50% Absorbed Energy Transition Temperature in V-Notch Charpy Impact Test vTrE (°C)
Weld metal	16Cr–8Ni LTT	KA36, 12 mm Thickness CO_2 GMAW Joints	–100
Weld metal	10Cr–10Ni LTT	KA36, 12 mm Thickness 80%Ar–20%CO_2 GMAW Joints	–200
Weld bond	Conventional Wire	KA36, 25 mm Thickness ESW Joints	–20 to 20
Weld bond	Conventional Wire	KD36, 25 mm Thickness SAW Joints	–20 to 20

2.2.4.3 Center-Notched Wide Plate Tensile Test for Fracture Toughness of LTT Weld Metals

To use 16%Cr–8%Ni LTT welding wire in ship structures, its weld metal must meet the Class-NK rules for steel ships [55] as specified in Tables 2.2.9 and 2.2.10.

TABLE 2.2.9

Impact Test and Tensile Test Requirements for Deposited Metal (Semi-Automatic Welding Consumables) [Class-NK Rules for Steel Ships]

Grade of Welding Consumable	Testing Temp. (°C)	Minimum Mean Absorbed Energy vE (J)	Yield Point (N/mm^2)	Tensile Strength (N/mm^2)	Elong. (%)
KSWL 2	–60	34	>345	440–610	>22
KSWL 3	–60	34	>375	490–660	>21

TABLE 2.2.10

Impact Test and Tensile Test Requirements for Butt Weld (Semi-Automatic Welding Consumables) [Class-NK Rules for Steel Ships]

Grade of Welding Consumable	Testing Temperature (°C)	Minimum Mean Absorbed Energy vE (J)		Tensile Strength (N/mm^2)
		Flat·Horizontal Overhead	Vertical Upward·Vertical Downward	
KSWL 2	–60	27	27	>440
KSWL 3	–60	27	27	>490

However, the V-notch Charpy absorbed energy of 16%Cr–8%Ni LTT weld metal with 100% CO_2 shielding gas and 10%Cr–10%Ni LTT weld metal welded with 80%Ar–20%CO_2 shielding gas does not fully satisfy the standard values given in Tables 2.2.9 and 2.2.10, as shown in Figures 2.2.23 and 2.2.25.

Here, center-notched wide plate tensile specimens with a test plate thickness of 12 mm and a test plate width of 200 mm were prepared. The fracture toughness values were determined, compared, and evaluated for the following weld metals:

1. 16%Cr–8%Ni LTT weld metal with 100% CO_2 shielding gas,
2. 10%Cr–10%Ni LTT weld metal with 80%Ar–20%CO_2 shielding gas, and
3. Weld metal using conventional carbon steel FCW (conventional wire).

2.2.4.3.1 Center-Notched Wide Plate Tensile Specimen

The test specimen was prepared by machining a butt-welded joint shown in Figure 2.2.22 into the shape and dimensions of the center-notched wide plate tensile specimen illustrated in Figure 2.2.27. A fatigue crack was introduced at the central notch tip under a gross stress range ($\Delta\sigma_{gross}$) of 40.0 N/mm^2. The number of stress cycles (N) for fatigue crack initiation ranged from 3.30×10^4 to 8.50×10^4. Due to the fatigue crack propagating within the weld metal, some variability was observed.

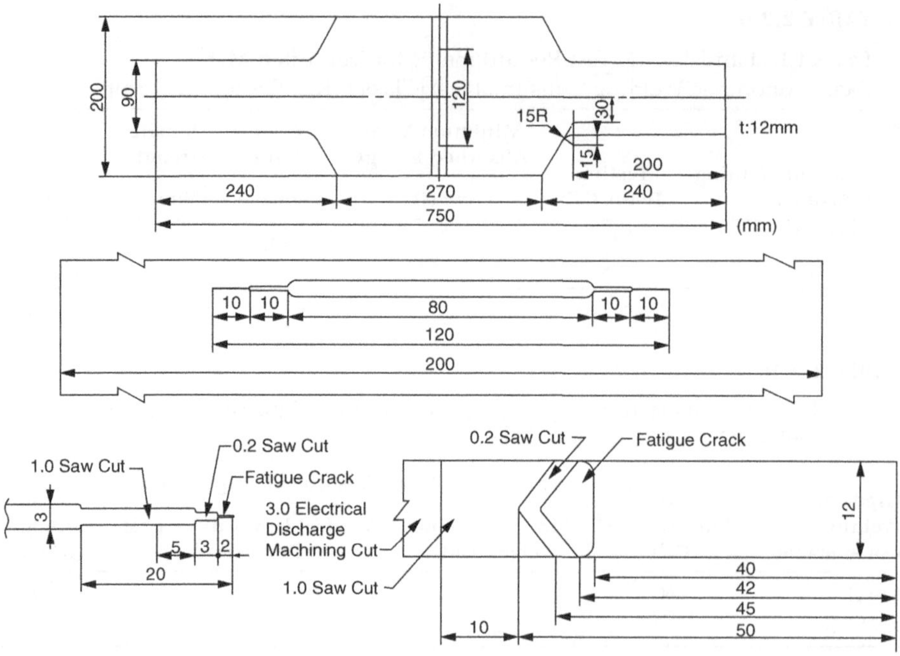

FIGURE 2.2.27
Shape of center-notched wide plate tensile specimen.

The preparation conditions for the initial fatigue crack are detailed in Table 2.2.11, which will be discussed later. The fatigue crack was generated using a 500 kN electro-hydraulic fatigue testing machine under the following conditions: room temperature, ambient air, axial tension, a constant load [stress ratio $R=0$], a load frequency of 5 Hz, and a sinusoidal load waveform. Additionally, the 16%Cr–8%Ni LTT butt joint specimens underwent aging heat treatment at 110°C for 12 hours before testing.

2.2.4.3.2 Results of the Center-Notched Wide Plate Tensile Test for LTT Weld Metal

A center-notched wide plate tensile specimen of the aforementioned butt joint was prepared, and low-temperature tensile tests for fracture toughness were conducted using a 2,000 kN Amsler-type universal testing machine [52,53].

The results of the low-temperature fracture toughness tests are presented in Table 2.2.11. The fracture toughness K_C values shown in Table 2.2.11 were calculated using Eq. (2.2.1) [56].

$$K_C = \sigma_g \cdot (\pi a)^{1/2} \cdot F(a/B) \quad (\mathrm{N/mm^{3/2}}) \qquad (2.2.1)$$

$$F(a/B) = [1 - 0.025(a/B)^2 + 0.06(a/B)^4] \times [\sec(\pi a/2B)]^{1/2}$$

TABLE 2.2.11

Results of Fracture Toughness Tests for LTT Weld Metal

	Specimen		Fatigue Crack						Fracture Stress		
	Width	Thickness	Stress[a] Range	Number of Cycles	Notch Length	Test Temperature		Fracture Load	Gross	Net	Fracture Toughness
Specimen Number	$2B$ (mm)	t (mm)	$\Delta\sigma_{gross}$ (N/mm²)	N (cycles)	$2a$ (mm)	T (°C)	$1/T_K$ (K^{-1})	P (kN)	σ_g (N/mm²)	σ_n (N/mm²)	K_C ($N/mm^{3/2}$)
LTT16/8-1[b]	200	12.2	40.0	3.80×10^4	124	−138	7.41×10^{-3}	463	190	499	3.53×10^3
LTT16/8-2[b]	200	12.2	40.0	3.30×10^4	121	−173	10.0×10^{-3}	245	100	254	1.81×10^3
LTT16/8-3[b]	200	12.1	40.0	3.30×10^4	121	−103	5.88×10^{-3}	475	196	497	3.54×10^3
LTT10/10-1	203	12.3	50.0	3.30×10^4	126	−176	10.3×10^{-3}	227	91	240	1.70×10^3
			40.0	2.20×10^4							
LTT10/10-2	201	12.4	40.0	5.50×10^4	123	−155	8.48×10^{-3}	277	110	284	2.03×10^3
LTT10/10-3	201	12.7	50.0	3.00×10^4	126	−47.4	4.43×10^{-3}	385	151	404	2.87×10^3
			40.0	1.70×10^4							
LTT10/10-4	201	12.0	40.0	8.50×10^4	117	−63.1	4.76×10^{-3}	402	167	399	2.90×10^3
LTT10/10-5	202	12.3	40.0	5.50×10^4	119	−128	6.90×10^{-3}	429	173	420	3.06×10^3
LTT10/10-6	203	12.3	40.0	5.50×10^4	124	−71.2	4.96×10^{-3}	400	160	412	2.95×10^3
LTT10/10-7	201	12.4	40.0	5.50×10^4	121	−139	7.46×10^{-3}	477	191	481	3.45×10^3
LTT10/10-8	203	12.1	40.0	5.50×10^4	125	−186	11.5×10^{-3}	236	96	250	1.79×10^3
Conv-1	201	12.5	40.0	7.00×10^4	121	−61.4	4.73×10^{-3}	500	199	500	3.59×10^3
Conv-2	203	11.9	40.0	8.50×10^4	126	−144	7.75×10^{-3}	192	80	210	1.50×10^3

[a] Stress ratio $R=0.0$.

[b] Heat treatment for aging: 110°C×12°C

σ_g: Gross stress (N/mm^2)

$2a$: Central through-notch length (mm)

$2B$: Specimen width (mm)

The relationship between gross fracture stress (σ_g) and temperature (T), as well as the relationship between net fracture stress (σ_n) and temperature (T), is shown in Figure 2.2.28. Furthermore, the relationship between fracture toughness (K_C value) and temperature ($1/T_K$) is illustrated in Figure 2.2.29.

Additionally, the dashed lines in Figures 2.2.28 and 2.2.29 represent accumulated historical data. These lines display the results of fracture toughness tests [57] conducted on center-notched wide plate tensile specimens similar to the one illustrated in Figure 2.2.27, with a width of 70 mm and a thickness of 12 mm. The tests were performed on high-tensile-strength (460 MPa grade for yield point) shipbuilding steel plates with electro-gas welding using general carbon steel wire. In the test specimen, the notch tip was located at the fusion line of the weld metal.

As indicated in Table 2.2.11, the fracture toughness values (K_C values) of the weld metals from two specimens, Conv-1 and Conv-2, using conventional wire can be plotted as dashed lines in Figures 2.2.28 and 2.2.29, confirming their consistency with the existing data.

As clearly demonstrated in Figures 2.2.28 and 2.2.29, both the 16%Cr–8%Ni LTT weld metal and the 10%Cr–10%Ni LTT weld metal exhibit fracture toughness values (K_C values) at low temperatures that surpass those of the

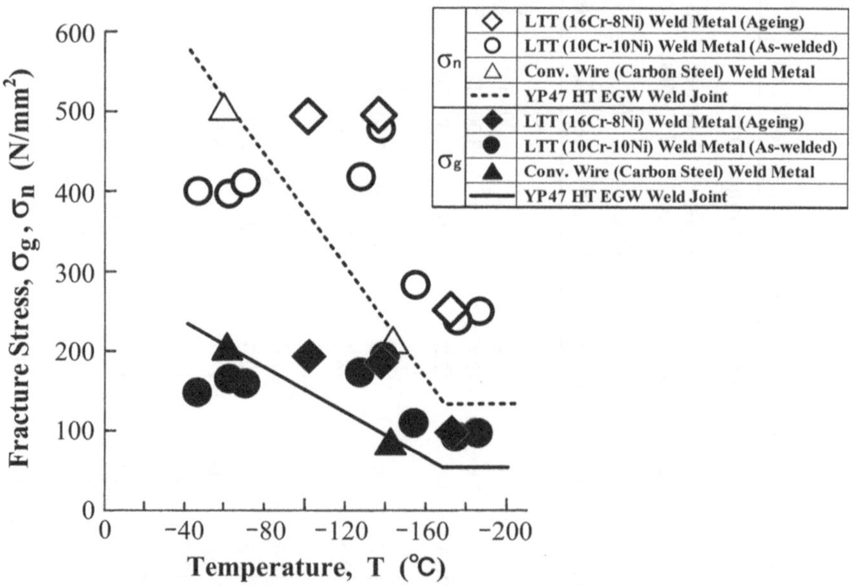

FIGURE 2.2.28
Relationship between fracture stress (σ_g, σ_n) and temperature.

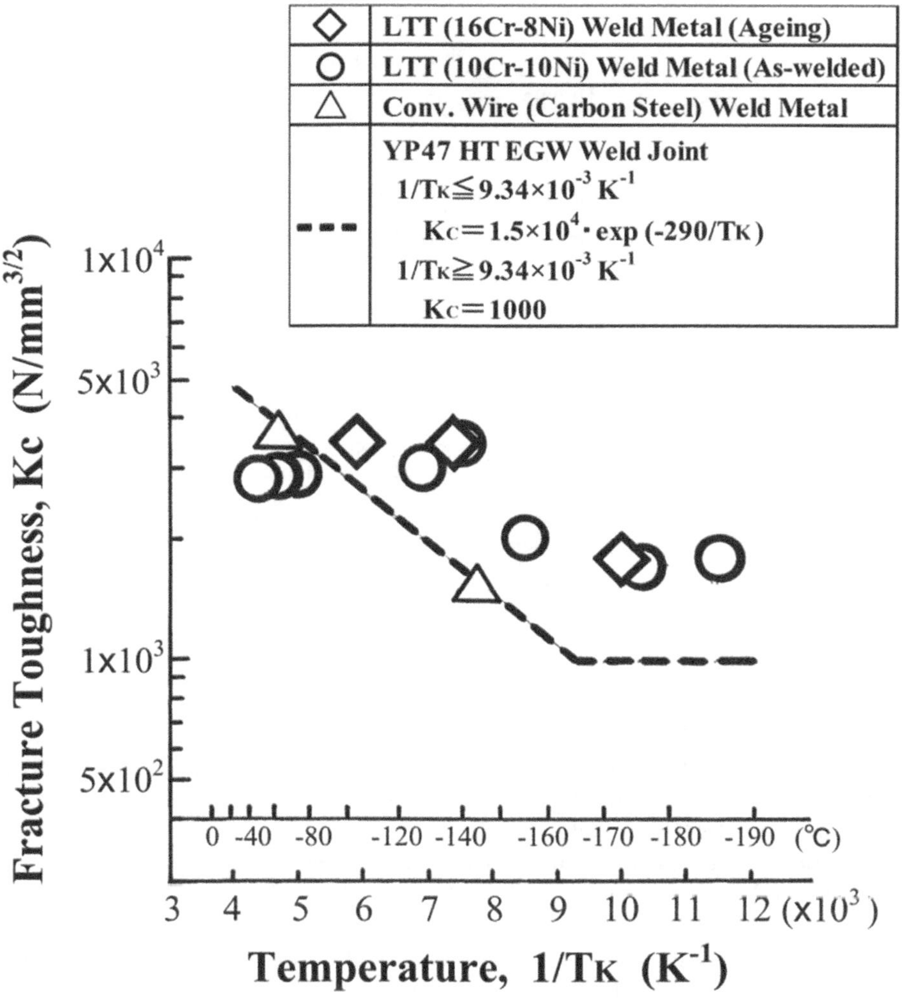

FIGURE 2.2.29
Results of fracture toughness tests for LTT weld metals (relation between K_C and temperature).

weld metal produced with conventional wire. This confirms that these materials possess superior low-temperature toughness.

2.2.4.3.3 Observation of Fracture Morphology Using Scanning Electron Microscopy

The fracture surfaces, observed by SEM, for the weld metal in conventional wire specimens from the aforementioned fracture toughness tests are presented in Figure 2.2.30. In Figure 2.2.30a, the Conv-1 specimen at a test temperature of −61°C exhibits ductile crack initiation from the fatigue crack tip,

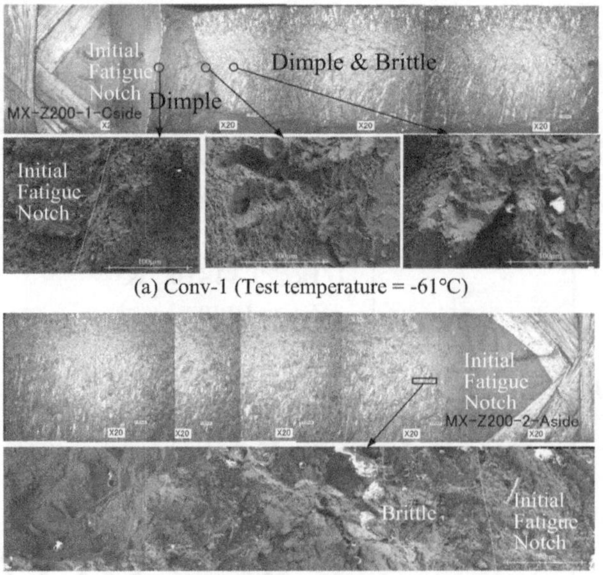

(a) Conv-1 (Test temperature = -61°C)

(b) Conv-2 (Test temperature = -144°C)

FIGURE 2.2.30
Fracture surface of conventional weld metal by SEM observation.

followed by unstable fracture originating from this ductile crack. Meanwhile, in Figure 2.2.30b, the Conv-2 specimen at a test temperature of –144°C shows brittle unstable crack initiation directly from the fatigue crack tip, leading to instantaneous brittle unstable fracture.

The SEM observation results of the test specimens LTT 16/8-1, 16/8-2, and 16/8-3 listed in Table 2.2.11 for the 16%Cr–8%Ni LTT weld metal are shown in Figure 2.2.31.

In Figure 2.2.31a for LTT 16/8-3 at a test temperature of –103°C and Figure 2.2.31b for LTT 16/8-1 at a test temperature of –138°C, ductile cracks initiated from the fatigue crack tip, leading to ductile unstable fracture. Additionally, in Figure 2.2.31c for LTT 16/8-2 at a test temperature of –173°C, a mixed ductile–brittle crack forms from the fatigue crack tip, resulting in ductile–brittle unstable fracture.

In contrast to the brittle fracture surface of Conv-2 at a test temperature of –144°C, the LTT 16/8 specimens exhibit a ductile fracture morphology even at much lower temperatures.

Figure 2.2.32 shows the SEM observation results of the test specimens LTT 10/10-1 to 10/10-8 for the 10%Cr–10%Ni LTT weld metal. In Figure 2.2.32a for LTT10/10-7 at a test temperature of –139°C, a ductile crack initiates from the fatigue crack tip, leading to ductile unstable fracture. This fracture morphology is identical to that of the aforementioned LTT16/8-1 at a test temperature of –138°C. Similarly, in Figure 2.2.30b for LTT10/10-1 at a test temperature of

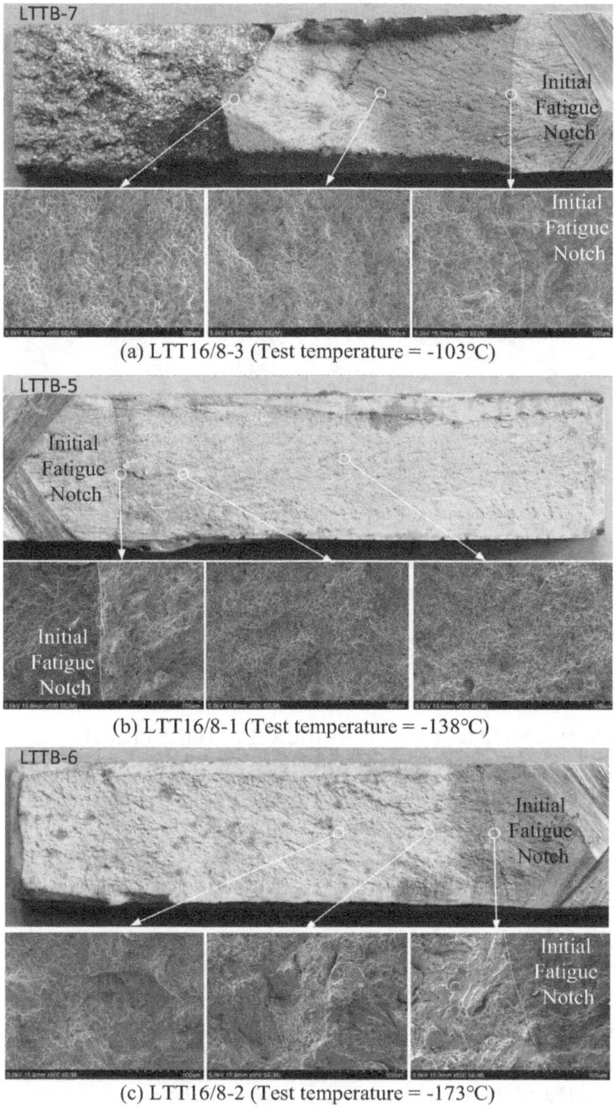

(a) LTT16/8-3 (Test temperature = -103°C)

(b) LTT16/8-1 (Test temperature = -138°C)

(c) LTT16/8-2 (Test temperature = -173°C)

FIGURE 2.2.31
Fracture surface of 16%Cr–8%Ni LTT weld metal by SEM observation.

–176°C, as with LTT16/8-2 at a test temperature of –173°C, a mixed ductile–brittle crack initiates from the fatigue crack tip, resulting in ductile–brittle unstable fracture.

From the fracture surface observations, the fracture toughness test results of the LTT weld metal and conventional carbon steel weld metal can be summarized as follows:

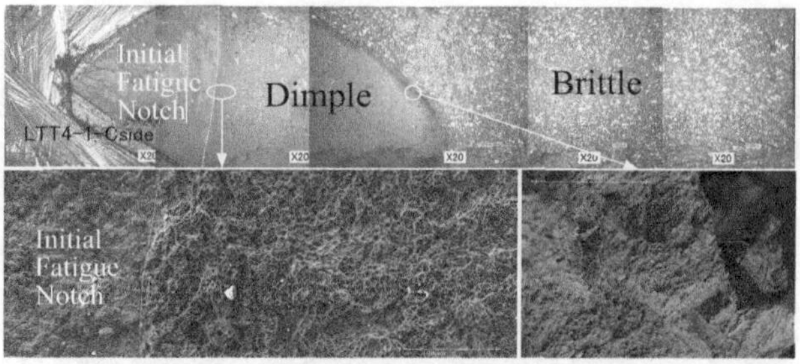

(a) LTT10/10-7 (Test temperature = -139°C)

(b) LTT10/10-1 (Test temperature = -176°C)

FIGURE 2.2.32
Fracture surface of 10%Cr–10%Ni LTT weld metal by SEM observation.

1. When the fracture test was conducted at temperatures around –60°C to –70°C, both the LTT weld metal and the conventional carbon steel weld metal initially exhibited ductile crack initiation at the fatigue crack tip, followed by unstable fracture originating from this ductile crack. In other words, the fracture modes of the two materials were entirely identical, and the fracture initiation was ductile in nature.
2. When the fracture test was performed at temperatures below approximately –140°C, the LTT weld metal, similar to the above case, first showed ductile crack initiation at the fatigue crack tip, which then led to unstable fracture. On the other hand, in the conventional carbon steel weld metal, a brittle unstable crack initiated directly at the fatigue crack tip, resulting in immediate brittle unstable fracture. The fracture modes of the LTT weld metal and the conventional carbon steel weld metal were completely different. The key distinction is that the LTT weld metal exhibited ductile fracture initiation, whereas the conventional carbon steel weld metal underwent brittle unstable fracture.

Based on the above, it can be concluded that even if fatigue cracks initiate and propagate in 16%Cr–8%Ni LTT weld metal welded with 100% CO_2 gas, these will not directly lead to brittle unstable fracture. This demonstrates that although the LTT weld metal exhibits a low upper shelf energy (approximately 30 J) in V-notch Charpy impact tests, its 50% absorbed energy transition temperature (vTrE) is very low, around –100°C. Unlike conventional carbon steel weld metals, the LTT weld metal possesses sufficient fracture toughness despite its low Charpy absorbed energy values.

2.2.5 Sea Water Corrosion in 16%Cr–8%Ni LTT Welded Joint

Section 2.1.4 describes the corrosion of metals in atmospheric environments, primarily focusing on structures such as bridges. This section focuses on marine applications, particularly the corrosion behavior of 16%Cr–8%Ni LTT welded joints in seawater. The evaluation results are based on a combined environmental cyclic corrosion test (alternating immersion test in artificial seawater), simulating the conditions in the upper part of ballast tanks [51].

2.2.5.1 Corrosion Potential of LTT Weld Metal in Artificial Seawater

A 490 MPa-class high-tensile-strength steel plate for ship hulls (KA36: thickness 12 mm) was used as the base metal, and weld metals were fabricated using a 16%Cr–8%Ni LTT wire (Kobelco: DW-168FP) and a conventional low-carbon steel wire (Kobelco: MX-Z200). The chemical compositions of the fabricated weld metals and the base metal are listed in Table 2.2.12. The surfaces of these joints and steel plates were ground and polished, and specimens with dimensions of 10 mm (width) × 15 mm (length) × 5 mm (thickness) including the weld metals and steel plates were extracted. An area of 5 mm × 10 mm was exposed as the potential evaluation surface, while the remaining areas were covered with corrosion-resistant tape to prepare the test specimens.

Figure 2.2.33 shows the electrochemical testing method. A saturated silver chloride reference electrode (Ag/AgCl (KCl)) was used. The test specimens of the 16%Cr–8%Ni LTT weld metal, conventional weld metal (MX-Z200), and base metal KA36 steel plate were immersed in a 3.5% NaCl artificial seawater solution. The corrosion potential was measured at a test temperature of 30°C for 3 hours.

The results are shown in Figure 2.2.34. While the corrosion potentials of the conventional weld metal and the base steel plate KA36 were nearly

TABLE 2.2.12

Chemical Compositions of Base Material and Weld Metals

		Chemical Composition (Mass%)								
Material	Wire	C	Si	Mn	P	S	Ni	Cr	Cu	Dilution Rate (%)
Base metal KA36	–	0.16	0.24	1.04	0.015	0.005	0.01	0.03	0.02	–
Weld metal	LTT (DW-168FP)	0.06	0.5	0.51	0.02	0.003	6.1	11.2	–	26
	Conventional (MX-Z200)	0.07	0.54	1.6	0.01	0.006	0.02	0.03	–	–

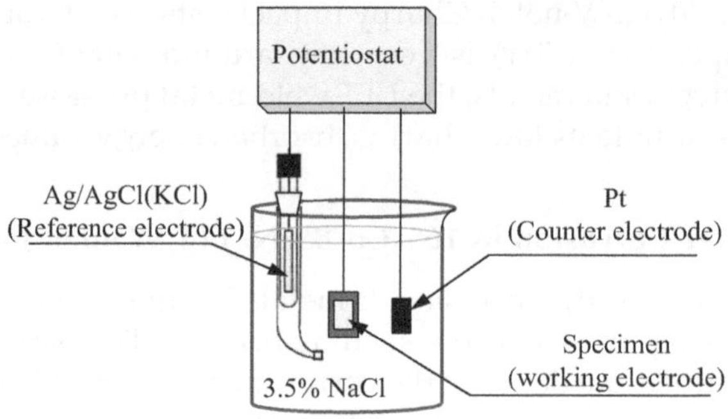

FIGURE 2.2.33
Electrochemical measurement for corrosion potential.

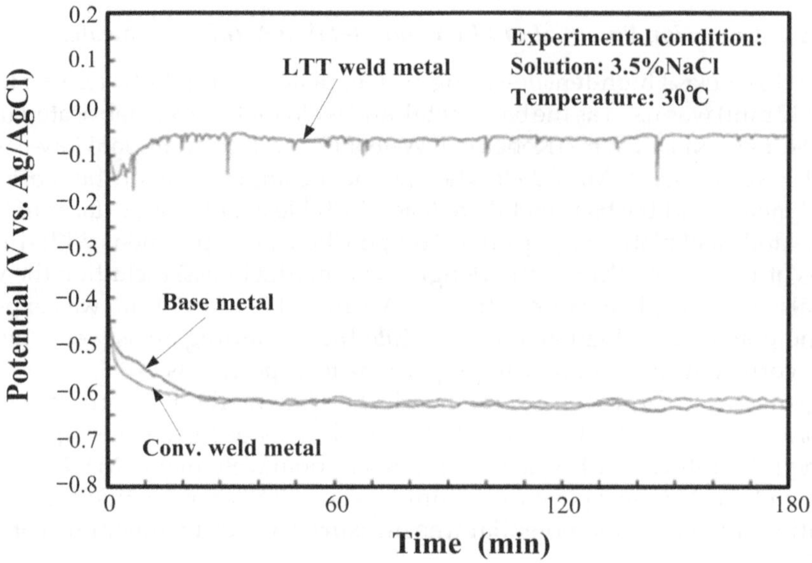

FIGURE 2.2.34
Measured corrosion potential of LTT weld metal, conventional weld metal, and base metal.

identical at approximately −0.66V, that of the 16%Cr–8%Ni LTT weld metal exhibited a higher potential of −0.07V. Therefore, corrosion tests were conducted focusing on two types of welded joints: conventional welded joints and 16%Cr–8%Ni LTT welded joints.

2.2.5.2 Procedure for Artificial Seawater Cyclic Corrosion Acceleration Test

The groove shape, dimensions, and photographs of the welded joints subjected to the corrosion acceleration test are shown in Figure 2.2.35. The chemical composition of the 16%Cr–8%Ni LTT weld metal diluted in this case is presented in Table 2.2.13. The base metal is KA36 steel plate with a thickness of 12 mm, as listed in Table 2.2.12, and the chemical composition of the conventional weld metal is nearly equivalent to that shown in Table 2.2.12.

The front and back surfaces of the joints fabricated in Figure 2.2.35b and c were ground and polished flat. Six pieces of test specimens with dimensions of 20 mm in width × 170 mm in length × 10 mm in thickness were extracted and subjected to the combined environmental cyclic test (corrosion acceleration test).

The most severe combined environmental cycle test (artificial seawater wet–dry cyclic test) simulating real corrosion conditions was conducted on conventional welded joint specimens (MX1-6) and 16%Cr–8%Ni LTT welded joint specimens (LTT 1-6).

The conditions of the combined environmental cycle test are shown in Figure 2.2.36. Cycle A and Cycle B were alternately repeated at 7-day intervals. These conditions are the same as the test conditions used in the joint research on corrosion-resistant steel by the Japan Iron and Steel Federation and the Nippon Kaiji Kyokai (ClassNK) [58].

The artificial seawater was prepared using ASTM D1141 substitute seawater equivalent, and the tests were conducted using the test apparatus shown in Figure 2.2.37 (Suga Test Instruments Co., Ltd.: CYP-90). Figure 2.2.37 also illustrates the arrangement of the test specimens. The notation "LTT" refers to LTT joints, while "MX" denotes standard joints. Six groups were formed, each consisting of a pair of standard joints and LTT joints. The specimens were rotated counterclockwise around the spray tower in pairs on a weekly

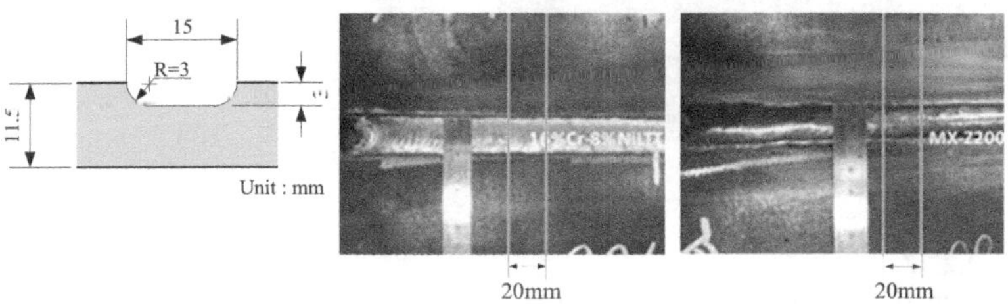

FIGURE 2.2.35
Measured corrosion potential of LTT weld metal, conventional weld metal, and base metal.

TABLE 2.2.13
Chemical Compositions of Diluted LTT Weld Metal

	Chemical Composition (Mass%)						
Weld Metal	**C**	**Si**	**Mn**	**P**	**S**	**Ni**	**Cr**
LTT (DW-168FP)	0.06	0.5	0.51	0.02	0.003	6.1	11.2

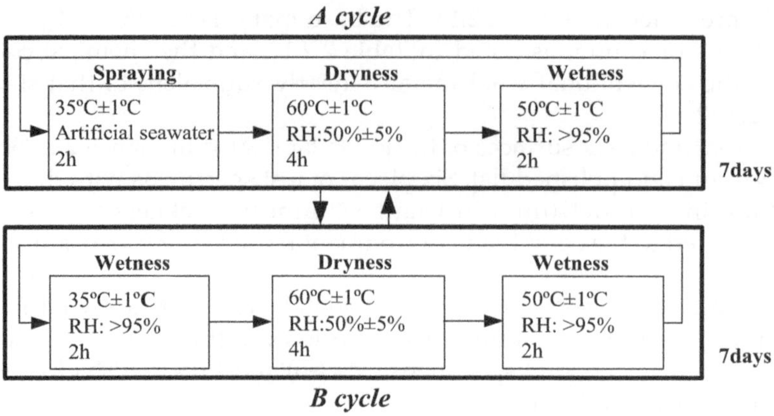

FIGURE 2.2.36
Accelerated cyclic corrosion tests under dry and wet conditions.

FIGURE 2.2.37
Arrangement of test specimens in the corrosion test chamber.

basis during the test. Each specimen underwent a corrosion test for 53 days in accordance with the specified cycle conditions.

2.2.5.3 Results of Accelerated Corrosion Test by Alternating Immersion in Artificial Seawater

Figure 2.2.38 shows the surface conditions at the end (53 days) of the test and after rust removal. For the conventional welded test specimens (MX), corrosion occurred across the entire surface, including the weld metal, in all

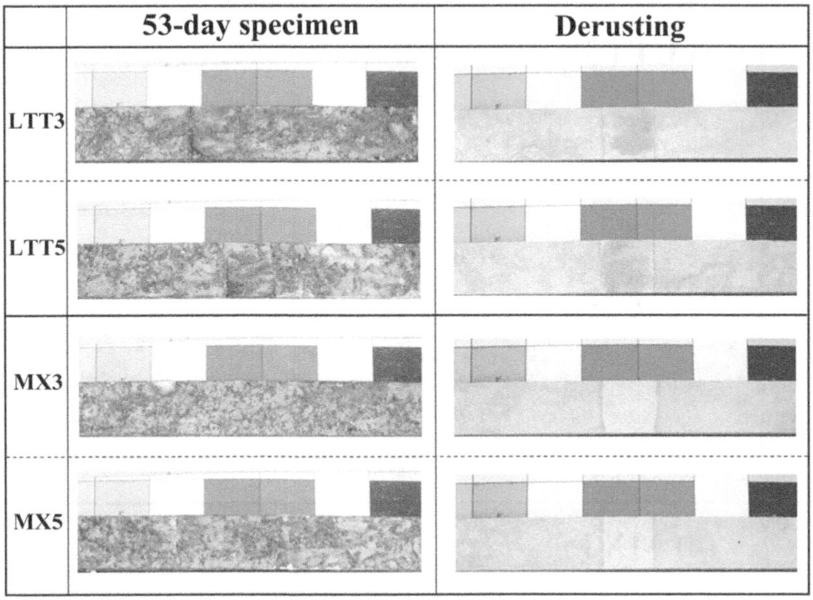

FIGURE 2.2.38
Appearance of specimens after cyclic corrosion testing.

specimens. In contrast, the weld metal area of the LTT welded test specimens (LTT) remained uncorroded and retained its luster.

Additionally, regarding the corrosion surface morphology, Figure 2.2.39 shows the profile (height distribution in the plate thickness direction) of the corroded surface of the test specimen measured using a 3D shape measuring instrument.

First, in the range of x=75–95 mm where the LTT weld metal (LTT3 and LTT5 test specimens) is present, almost no corrosion occurs, while the base metal section exhibits nearly uniform corrosion of approximately 0.1 mm across the entire surface from the LTT surface.

In other words, the corrosion surface gradient at the LTT/base metal boundary is steep, but no localized corrosion concentrated at the interface, suggesting that galvanic corrosion was not observed.

Next, in the case of conventional weld metal (MX3 and MX5 test specimens) in Figure 2.2.39, nearly uniform general corrosion occurred in both the conventional weld metal zone and the base metal, with no difference in corrosion behavior observed between them.

Based on the above, composite cyclic corrosion tests simulating the upper environment of a ballast tank were conducted on unpainted onboard weld test specimens using LTT and conventional weld materials. The results showed that in the LTT specimens, almost no corrosion occurred in the LTT weld metal, and although the corrosion surface gradient at the boundary

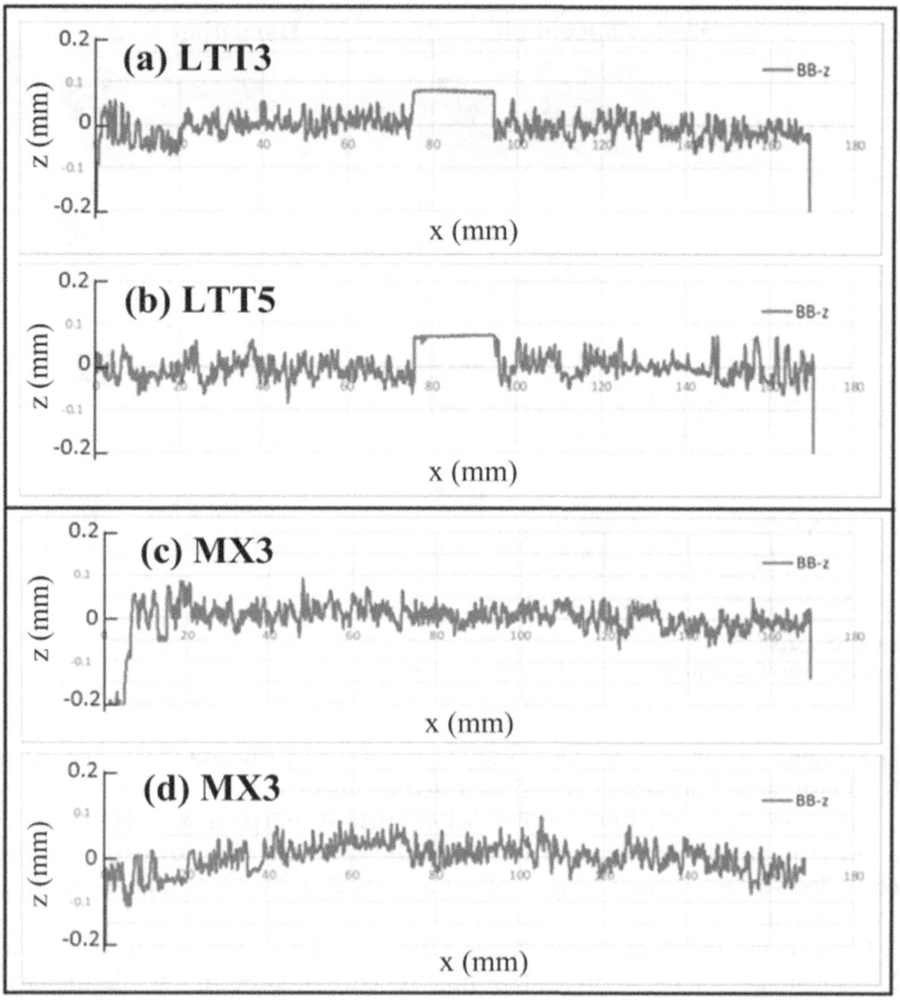

FIGURE 2.2.39
Surface profile of welded joints after the cyclic corrosion test.

between the LTT material and the base metal was steep, localized corrosion at the interface was not observed. In the conventional specimens, nearly uniform general corrosion occurred in both the weld metal and the base metal, with no discernible difference in corrosion behavior between the weld metal and the base metal.

These results were consistent with the atmospheric exposure test results in Section 2.1.4, leading to the interference that the absence of galvanic corrosion was due to the presence of a rust layer.

References

1. S. Zenitani, N. Hayakawa, J. Yamamoto, K. Hiraoka, Y. Morikage, T. Kubo, K. Yasuda, and K. Amano, Development of new low transformation temperature welding consumable to prevent cold cracking in high strength steel welds. *Science and Technology of Welding and Joining*, 12(6) (2007), 516–522.
2. J. Yamamoto, S. Meguro, Y. Muramatsu, N. Hayakawa, and K. Hiraoka, Analysis of martensite transformation behavior in welded joint of low transformation-temperature materials. *Welding International*, 23 (2009), 411–421.
3. H. Qiu, J. Qi, F. Yin, and K. Hiraoka, Determination of parameters for fitting the dilatation curve of austenite–martensite transformation in Cr–Ni steels. *ISIJ International*, 49(1) (2009), 146–148.
4. C.L. Magee, *Phase Transformation*. American Society for Metals, Metals Park, Ohio (1970), 115–156.
5. D.P. Koistinen and R.E. Marburger, A general equation prescribing the extent of the austenite-martensite transformation in pure iron-carbon alloys and plain carbon steels. *Acta Metallurgica Sinica*, 7 (1959), 59.
6. M. De Meyer, J. Mahieu, and B.C. De Cooman, Empirical microstructure prediction method for combined intercritical annealing and bainitic transformation of TRIP steel. *Materials Science and Technology*, 18 (2002), 1121.
7. K. Hiraoka, N. Ma, Z. Feng, S. Kano, and M. Nagami, Design of 16%Cr–8%Ni LTT welding material and its weldability evaluation. *Preprints of the National Meeting of the Japan Welding Society*, 112 (2023), 170–171 (in Japanese).
8. K. Hiraoka, F. Yin, S. Meguro, and H. Qiu, Martensitic transformation and hydrogen evolution behavior in weld metal of Ni-Cr steel. In *NEDO, P07005, 1st Symposium Proceedings*, Japan (2009), 39–40 (in Japanese).
9. H. Murata, N. Kato, and H. Tamura, Effect of transformation on residual stress in welding. *Quarterly Journal of the Japan Welding Society*, 11(4) (1993), 545–550 (in Japanese).
10. S. Zenitani, N. Hayakawa, J. Yamamoto, K. Hiraoka, C. Shiga, Y. Morikage, T. Kubo, and K. Yasuda, Prevention of cold cracking in high strength steel welds by applying newly developed low transformation-temperature welding consumables. In *6th International Trends in Welding Research Conference Proceedings*, ASM International, Pine Mountain, USA, April 15–19 (2002–2003), 569–574.
11. V.P. Kujanpaa, N. Suutala, T. Takalo, and T. Moisio, Correlation between solidification cracking and microstructure in austenitic and austenitic-ferritic stainless steel welds. *Welding Research International*, 9(2) (1979), 55–75.
12. N. Suutala, T. Takalo, and T. Moisio, ferritic-austenitic solidification mode in austenitic stainless steel welds. *Metallurgical Transactions A*, 11A (1980), 717–725.
13. K. Nishimoto, Welding of stainless steels. *Journal of the Japan Welding Society*, 66(3) (1997), 156–173 (in Japanese).
14. R.D. Stout, R. Vasudevan, and A.W. Pense, A field weldability test for pipeline steels. *Welding Journals*, 55(4) (1976), 89s–94s.
15. N. Yurioka, H. Suzuki, S. Ohshita, and S. Saito, Determination of necessary preheating temperature in steel welding. *Welding Journals*, 62(6) (1983), 147s–153s.
16. T. Kasuya and N. Yurioka, Carbon equivalent and multiplying factor for hardenability of steel. *Welding Journals*, 72(6) (1993), 263s–268s.

17. Y. Ito and K. Bessyo, Cracking parameter of high strength steels related to heat affected zone cracking (report 2). *Journal of the Japan Welding Society*, 38(10) (1969), 1134–1144 (in Japanese).
18. K. Satoh and S. Matsui, Process of contraction during first pass welding and cracking (part 3). *Journal of the Japan Welding Society*, 35(6) (1966), 413–420 (in Japanese).
19. F. Matsuda, H. Nakagawa, K. Shinozaki, and Y. Nishio, Relaxation phenomenon of the restraint stress due to transformation expansion in weld zone of HY-steel. *Quarterly Journal of the Japan Welding Society*, 5(2) (1987), 250–255 (in Japanese).
20. A. Kromm and T. Kannengiesser, Characterizing phase transformations of different LTT alloys and their effect on residual stresses and cold cracking. Document IX-2311-09 (IX-H-708-09).
21. K. Shinozaki, L. Ke, and T. H. North, Hydrogen cracking in duplex stainless steel weld metal. *Welding Journals*, 71(11) (1992), 387s–396s.
22. M. Koibuchi and T. Yokota, Underwater wet welding with Ni, Fe-Ni and stainless steel electrodes. *Journal of the Japan Welding Society*, 50(5) (1981), 489–495 (in Japanese).
23. M. Mochizuki, K. Mikami, K. Saida, and T. Kasuya, Hydrogen diffusion-accumulation-discharge behavior of ultra-high strength weld metal with retained austenite. In *NEDO, P07005, 2nd Symposium Proceedings*, Japan (2012), pp. 31–32 (in Japanese).
24. Y. Mikami, N. Kawabe, N. Ishikawa, and M. Mochizuki, Evolution of cold cracking in high-strength steel weld metal based on local critical conditions incorporating stress and diffusible hydrogen distributions. *Quarterly Journal of the Japan Welding Society*, 34(2) (2016), 67–80 (in Japanese).
25. F. Gunic, N. Hayakawa, K. Hiraoka, and M. Katabami, Improvement of mechanical properties of LTT weld metal. *Preprints of the National Meeting of the Japan Welding Society*, 76 (2005), 10–11.
26. S. Terashima and H.K.K.D.H. Bhadeshia, Size distribution of oxides and toughness of steel weld metals. *Science and Technology of Welding and Joining*, 11(5) (2006), 580–582.
27. S. Zenitani, T. Nakamura, K. Hiraoka, and K. Shinozaki, Ar-MIG arc behavior in local addition of small amount of oxygen into shielding gas. *Quarterly Journal of the Japan Welding Society*, 25(1) (2007), 187–195 (in Japanese).
28. M. Tanaka, T. Tamaki, K. Nakata, T. Ohnawa, and T. Ueyama, Development of plasma GMA welding method (Report 1). *Preprints of the National Meeting of the Japan Welding Society*, 80 (2007), 16–17 (in Japanese).
29. T. Nakamura and K. Hiraoka, *Development of Stable MIG Welding in Pure Ar Gas by New-Hybrid Wire*, IIW Doc. 212-1080-05, IIW (2005).
30. H. Qiu, F. Yin, Z. Hong, and K. Hiraoka, Measurement and verification of stress induced phase transformation in high Ni-Cr clean weld metal. In *NEDO, P07005, 1st Symposium Proceedings*, Japan (2009), pp. 43–44 (in Japanese).
31. R.D. Strommen, H. Horn, and K.R. Wold, FSM a unique method for monitoring corrosion, pitting, erosion and cracking. NACE Corrosion Paper 7 (1992).
32. C. Shiga, H. Yasuda, Y.-C. Kim, S.-H. Lee, and K. Hiraoka, Residual stress reduction and fatigue strength improvement of high strength steel welded joints. In *NEDO, P07005, 1st Symposium Proceedings*, Japan (2009), pp. 47–48 (in Japanese).
33. K. Oku, Y.-C. Kim, K. Arita, and H. Horikawa, Crack propagation behavior monitored by FSM. In *59th Japan Society of Civil Engineers Annual Meeting*, Japan (2004–2009) (in Japanese).

34. A. Ohta, O. Watanabe, K. Matsuoka, Y. Maeda, N. Suzuki, and T. Kubo, Fatigue strength improvement of box welds by using low transformation temperature welding material. *Quarterly Journal of the Japan Welding Society*, 19(2) (2001), 373–376 (in Japanese).
35. S. Zenitani, T. Nishimura, N. Hayakawa, and K. Hiraoka, Atmospheric corrosion behavior of high strength steel weld joints formed by low transformation-temperature welding consumables. *Quarterly Journal of the Japan Welding Society*, 24(4) (2006), 291–298 (in Japanese).
36. M. Kowaka, *Metal Corrosion Damage and Corrosion Protection*, Materials Science & Technology, AGNE, Lviv, Ukraine (1983), pp. 54–57 (in Japanese).
37. A. Tamada and M. Honda, Corrosion protection of steel structure in an arctic ocean by stainless steel. *Zairyo-to-Kankyo*, 40 (1991), 583–588 (in Japanese).
38. R. Francis, Galvanic corrosion of high alloy stainless steels in sea water. *British Corrosion Journal*, 29(1) (1994), 53–57.
39. H. Kimoto, Study of galvanic corrosion behavior for aluminized steel in seawater. *Zairyo-to-Kankyo*, 48 (1999), 432–438 (in Japanese).
40. A. Ohta, O. Watanabe, K. Matsuoka, C. Shiga, S. Nishijima, Y. Maeda, N. Suzuki, and T. Kubo, Fatigue strength improvement by using newly developed low transformation temperature welding material. *Welding in the World*, 43 (1999), 38–42.
41. T. Matsuzaki, K. Hiraoka, Z. Feng, N. Ma, H. Murakawa, S. Kano, K. Okada, S. Kimura, C. Shiga, and H. Yajima, New LTT welding material and all-position repair welding with elongated bead for super-long fatigue life of boxing fillet joints. *Journal of the Japan Society of Naval Architects and Ocean Engineers*, 32(12) (2020), 153–161 (in Japanese).
42. S. Kano, M. Nagami, N. Ma, and K. Hiraoka, Development of LTT welding consumables for all position welding with excellent crack resistance. *Conference proceedings, the Japan Society of Naval Architects and Ocean Engineers*, 34(11) (2023), 193–194 (in Japanese).
43. Z. Feng, N. Ma, K. Hiraoka, Y. Komizo, S. Kano, and M. Nagami, Development of 16Cr8Ni low transformation temperature welding material for optimal characteristics under various dilutions due to all repair welding positions. *Science and Technology of Welding and Joining*, 28(4) (2023), 305–313.
44. M. Yonemura, T. Osuki, H. Terasaki, Y. Komizo, M. Sato, and H. Toyokawa, Two-dimensional time-resolved X-ray diffraction study of directional solidification in steels. *Materials Transactions*, 47(9) (2006), 2292–2298.
45. Y. Komizo and H. Terasaki, In-situ observation for solidification and phase transformation during welding. *Journal of the Japan Society for Heat Treatment*, 54(1) (2014), 12–21 (in Japanese).
46. Y. Morikage, T. Kubo, K. Yasuda, K. Amano, K. Hiraoka, A. Ohta, and C. Shiga, Improvement of cold cracking susceptibility using low transformation temperature welding consumables. *Journal of Web Semantics*, 67 (2000), 510–511 (in Japanese).
47. ClassNK, *Guidance for the Survey and Construction of Ship Part M Chapter M4, Tests for Fillet Weld Joints M4.3* (2024), pp. 18–19.
48. M. Miyata and R. Suzuki, Welding process and consumables aimed at improving fatigue strength of joints. *Kobe Steel Engineering Reports*, 65 (2015), 16–20 (in Japanese).

49. S. Wu, D. Wang, Z. Zhan, C. Li, X. Liu, X. Meng, Z. Feng, and X. Di, Mechanical properties of low-transformation-temperature weld metals after low-temperature postweld heat treatment. *Science and Technology of Welding and Joining*, 24 (2019), 112–120.
50. S. Wu, et al., Strength-toughness improvement of martensite-austenite dual phase deposited metals after austenite reversed treatment with short holding time. *Materials Science and Engineering: A*, 755 (2019), 57–65.
51. T. Matsuzaki, N. Ma, K. Hiraoka, H. Murakawa, N. Osawa, T. Takeuchi, K. Okada, C. Shiga, and H. Yajima, Fatigue life extension of box welded joints by repaired or reinforced LTT elongated bead. *Conference proceedings, the Japan Society of Naval Architects and Ocean Engineers*, 37 (2023), 107–112 (in Japanese).
52. T. Matsuzaki, H. Murakawa, N. Ma, S. Tsutsumi, K. Hiraoka, N. Osawa, K. Okada, T. Tanino, C. Shiga, and H. Yajima, A study on fatigue life improvement of welded joints using elongated bead welding with low transformation temperature welding material. *Journal of High Pressure Institute of Japan*, 58(1) (2020), 22–38 (in Japanese).
53. T. Matsuzaki, N. Ma, K. Hiraoka, H. Murakawa, Z. Feng, K. Okada, S. Kimura, S. Kano, C. Shiga, and H. Yajima, A study on fatigue life improvement of welded joints in all positions and fracture toughness of weld metal for CO_2 gas Arc elongated bead welding method with low transformation temperature welding material. *Journal of High Pressure Institute of Japan*, 58(6) (2020), 315–328 (in Japanese).
54. R. Nagamoto, M. Ushijima, A. Kawamura, H. Yajima, and M. Tada, Fracture toughness of 50 kgf/mm^2 class new-CR process steel plates. *Journal of the Society of Naval Architects of Japan*, 152(11) (1982), 405–416 (in Japanese).
55. ClassNK, *Rules for the Survey and Construction of Steel Ships*, Part M, Chapter 6 (2024), pp. 79–82.
56. L. Noels, *Stress Intensity Factors Handbook, Volume 1, Committee on Fracture Mechanics, The Society of Materials Science*. Pergamon Press, Japan (1987), p. 3.
57. H. Yajima, E. Watanabe, Z.M. Jia, K. Yoshimoto, T. Ishikawa, and Y. Funatsu, Study on fracture toughness of welded joints for heavy-thick steel plates bycentre-notched small size specimen. *Welding in the World*, 55(05) (2011), 84–92.
58. ClassNK and JISF, *Result report of Class NK R&D Project* (2016). https://www.classnk.or.jp/classnk-rd/assets/pdf/report/2016/13-08_1.pdf

3

Different LTT Welding Materials and Their Induced Residual Stress in Various Welded Joints

3.1 Numerical Modeling for Residual Stress Analysis

3.1.1 Finite Element Analysis for Residual Stress

The finite element (FE) method (FEM) has been widely employed in the analysis of welding-induced thermal fields and stress/strain fields [1]. There are many types of elements, such as solid, shell, and beam elements, in which solid elements are often used to model the geometry of welded joints. Figure 3.1.1 shows an example of solid elements in analyzing welding stresses in a fillet-welded joint. To reproduce the high temperature gradient and the strong nonlinearity of materials at high temperatures,

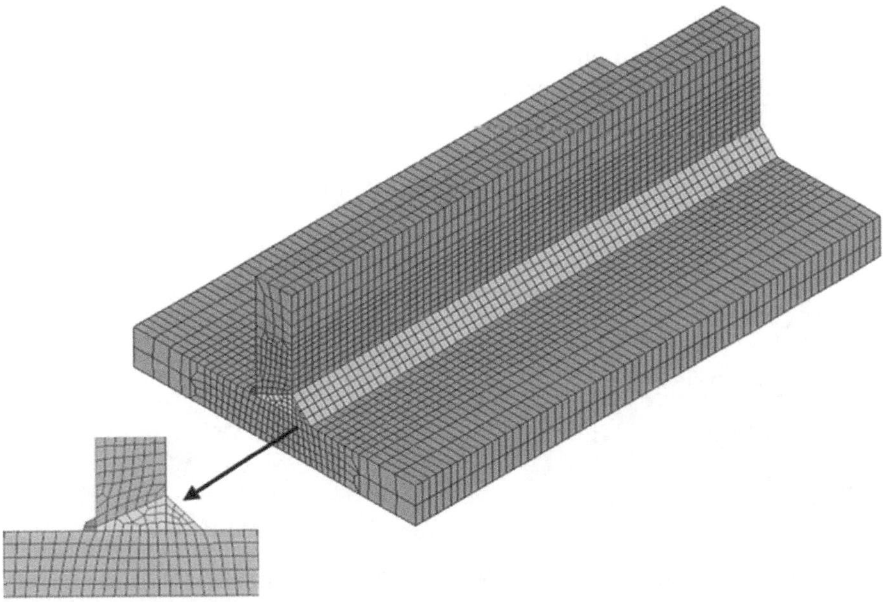

FIGURE 3.1.1
Example of finite elements for welding thermal stress/strain analysis.

DOI: 10.1201/9781003601845-3

small-sized elements in the welded metal (WM), and heat-affected zone (HAZ) are always used. To achieve good analysis efficiency, the base metal (BM), which is subjected to lower welding-induced temperatures, is meshed with relatively larger-sized elements. In this example, the element size in WM and HAZ ranges from 0.5 to 1.0 mm, while the maximum element size in BM is 10 mm.

3.1.2 Welding Heat Source Model for Thermal Conduction Analysis

To predict the welding-induced temperature field, the welding energy and its distribution entering the molten zone must be described by a mathematical equation called as the heat source model. Usually, a simple uniform distribution or a Gaussian distribution in the zone molten by the welding heat, as shown in Figure 3.1.2a and b, respectively, is used in the finite element analysis (FEA) for welding thermal conduction [1,2].

3.1.3 Strains for Analysis of Welding Stress in LTT Materials

3.1.3.1 Total Strain

Since the welding thermal stress and strain phenomena are strongly nonlinear problems, an incremental analysis approach is generally employed. Therefore, the total strain is calculated by using the following equation by adding a new increment to the strain obtained in the previous step.

$$\{\varepsilon\}_{t+dt} = \{\varepsilon\}_t + \{d\varepsilon\} \tag{3.1.1}$$

The total strain increment $\{d\varepsilon\}$ consists of the increments of elastic strain $\{d\varepsilon^e\}$, plastic strain $\{d\varepsilon^p\}$, thermal strain $\{d\varepsilon^T\}$, and phase transformation strain $\{d\varepsilon^{ph}\}$.

$$\{d\varepsilon\} = \{d\varepsilon^e\} + \{d\varepsilon^T\} + \{d\varepsilon^p\} + \{d\varepsilon^{ph}\} \tag{3.1.2}$$

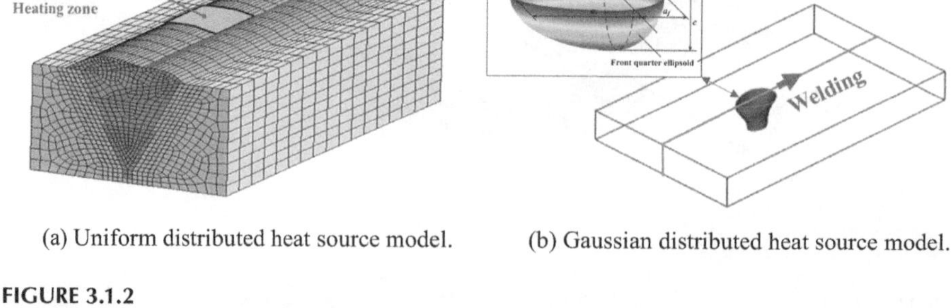

(a) Uniform distributed heat source model. (b) Gaussian distributed heat source model.

FIGURE 3.1.2
Welding heat source models. (a) Uniformly distributed heat source model. (b) Gaussian distributed heat source model.

The total strain increment $\{d\varepsilon\}$ at the integration points of an element can be computed from the nodal displacement increment $\{du\}$ using the displacement–strain corresponding matrix [B].

$$\{d\varepsilon\} = [\mathrm{B}]\{du\} \tag{3.1.3}$$

3.1.3.2 Elastic Strain

The elastic strain $\{d\varepsilon^{e}\}$ and its increment $\{\Delta\varepsilon^{e}\}$ are linearly proportional to the stress $\{\sigma\}$ and $\{\Delta\sigma\}$, respectively, following Hooke's law.

$$\{\varepsilon^{e}\} = [\mathrm{C}]\{\sigma\} \tag{3.1.4a}$$

$$\{\Delta\varepsilon^{e}\} = [\mathrm{C}]\{\Delta\sigma\} + [\Delta\mathrm{C}]\{\sigma\} \tag{3.1.4b}$$

where [C] is the compliance matrix of the material, defined by Young's modulus E and Poisson's ratio ν.

3.1.3.3 Plastic Strain

The plastic strain $\{\varepsilon^{p}\}$ and its increment $\{\Delta\varepsilon^{p}\}$ are produced during the loading state with plastic deformation, which depend not only on the current stress state but also on the loading history. Therefore, the total plastic strain must be calculated from its increment.

$$\{\Delta\varepsilon^{p}\} = \Delta\lambda \frac{\partial\bar{\sigma}}{\{\partial\sigma\}} \tag{3.1.5}$$

$$\{\varepsilon^{p}\}_{t+dt} = \{\varepsilon^{p}\}_{t} + \{\Delta\varepsilon^{p}\} \tag{3.1.6}$$

where $\Delta\lambda$ and $\bar{\sigma}$ are plasticity loading variable and equivalent stress, respectively, which will be discussed later.

3.1.3.4 Thermal Strain

The thermal strain $\{\varepsilon^{T}\}$ and its increment $\{\Delta\varepsilon^{T}\}$ are generally given by the following equations using the average thermal expansion coefficient α_0 or instantaneous thermal coefficient α.

$$\{\varepsilon^{T}\} = \{\alpha_0, \alpha_0, \alpha_0, 0, 0, 0\}^{T}(T - T_{ref}) \tag{3.1.7a}$$

$$\{\Delta\varepsilon^{T}\} = \{\alpha, \alpha, \alpha, 0, 0, 0\}^{T}\Delta T \tag{3.1.7b}$$

where T, ΔT, and T_{ref} are the current temperature, temperature increment, and reference temperature for defining the average thermal expansion coefficient, respectively.

The relationship between the average thermal expansion coefficient α_0 and the instantaneous thermal coefficient α can be expressed by the following equation and schematically demonstrated in Figure 3.1.3 [1]:

$$\alpha(T) = \alpha_0(T) + \frac{d\alpha_0(T)}{dT}(T - T_{\text{ref}}) \tag{3.1.8}$$

In addition to the thermal strain due to thermal expansion, the phase transformation-induced volume change, as well as the changes in Young's modulus and yield stress with temperature, will also produce additional thermal strain, which will be described later.

Figure 3.1.4 shows the measured expansion and contraction history of LTT welding material 16%Cr–8%Ni deposited on KA36 steel plate by formaster test during fast heating and cooling processes (20°C/s) between room temperature and 1,300°C [3]. The contraction between Ac_1 and Ac_3 temperatures during heating can be observed, which is induced by phase transformation from martensite to austenite. During fast cooling, the expansion can be observed between martensite start (Ms) and martensite finish (Mf) temperatures due to the phase transformation from austenite to martensite. The type of phase transformation-induced contraction and expansion presents a volume change $\{\Delta\varepsilon^{\text{vol}}\}$ similar to thermal contraction and expansion.

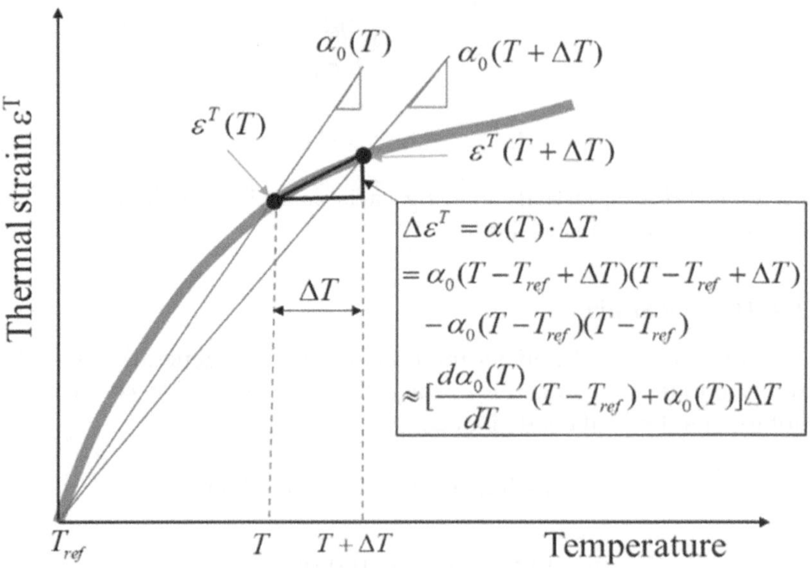

FIGURE 3.1.3
Average and instantaneous thermal expansion coefficients.

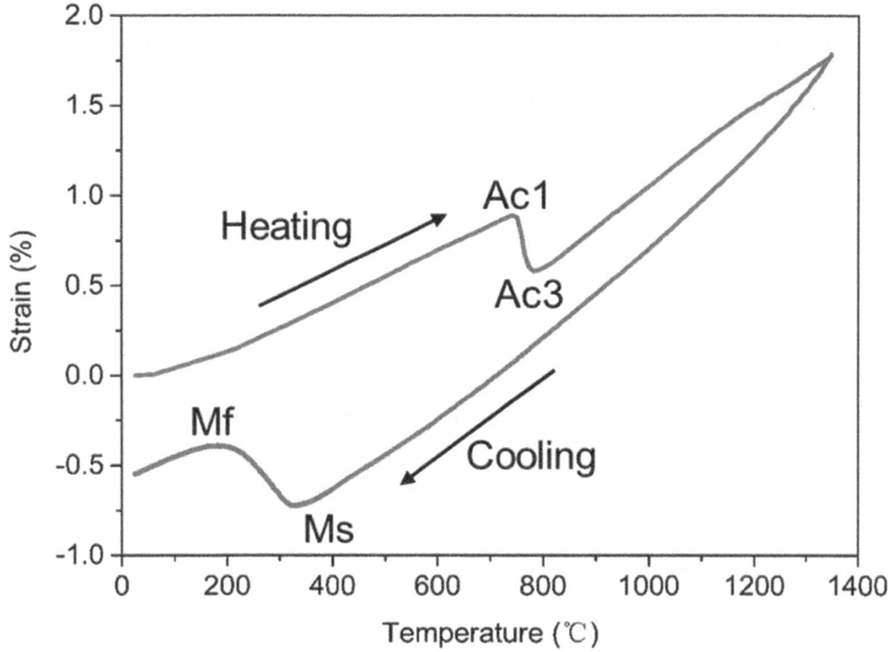

FIGURE 3.1.4
Measured thermal strain of 16%Cr8%Ni weld metal during heating and cooling.

3.1.3.5 Phase Transformation Strain

In addition to the volume change produced by phase transformation, there exists another plastic strain $\{\Delta\varepsilon^{tp}\}$ without volume change. Therefore, the phase transformation strain increment includes the following two components:

$$\{\Delta\varepsilon^{ph}\} = \{\Delta\varepsilon^{vol}\} + \{\Delta\varepsilon^{tp}\} \tag{3.1.9}$$

The phase transformation-induced volume strain $\{\Delta\varepsilon^{vol}\}$ can be treated as the equivalent thermal strain.

$$\{\Delta\varepsilon^{T}_{eq}\} = \{\Delta\varepsilon^{T}\} + \{\Delta\varepsilon^{vol}\} \tag{3.1.10}$$

The other term, namely phase transformation-induced non-volumetric strain $\{\Delta\varepsilon^{tp}\}$ from austenite to martensite during cooling, was formulated by Inoue [4,5] and Leblond [6] using the following equation, which is proportional to the stress deviation $\{\sigma'\}$:

$$\{\Delta\varepsilon^{tp}\} = 3K(1 - f_M)\Delta f_M\{\sigma'\} \tag{3.1.11}$$

where f_M and Δf_M are the fraction of martensite and its increment, respectively; and K is the coefficient of transformation plasticity, which can be identified using experimental data.

Till now, the coefficients of transformation plasticity (K parameter) of steels measured by experiment are very limited. It was reported in references [7,8].

The martensite fraction increment Δf_M is given by the Koistinen–Marburger (K–M) equation [9].

$$f_M = 1 - e^{-m(M_s - T)}, \quad \Delta f_M = m \cdot e^{-m(M_s - T)} \Delta T, \quad M_f \le T \le M_s \tag{3.1.12}$$

It can be seen that Δf_M is the function of temperature T but is not dependent on time. m is a parameter obtained by fitting experimental data. Therefore, it belongs to non-diffusive phase transformation.

3.1.4 Stress Analysis Based on Thermal–Elastic–Plastic Theory

The thermal stress $\{\sigma\}$ due to thermal loading during welding and its cooling processes can be calculated from the elastic strain $\{\varepsilon^e\}$ using Hooke's law:

$$\{\sigma\} = [D^e]\{\varepsilon^e\} \tag{3.1.13}$$

where $[D^e]$ is the material elastic matrix, which is the inverse matrix of the material compliance matrix [C] used in Eq. (3.1.4).

Because of the high nonlinearity of welding-induced thermal stress, it is usually updated in incremental form as follows:

$$\{\Delta\sigma\} = [D^e]\{\Delta\varepsilon^e\} + [\Delta D^e]\{\varepsilon^e\} \tag{3.1.14}$$

$$\{\sigma\}_{t+dt} = \{\sigma\}_t + \{\Delta\sigma\} \tag{3.1.15}$$

To accurately calculate the stress increment $\{\Delta\sigma\}$ in LTT materials, the yield strength changes with temperature during both heating and cooling processes, as schematically shown in Figure 3.1.5 are particularly important [1,10], besides the phase transformation-induced plasticity and the temperature dependence of Young's modulus and thermal expansion described above.

3.2 Residual Stress in a Simple Bar Model

To understand the generation mechanism of compressive residual stress in the LTT WM, a simple bar model shown in Figure 3.2.1 is employed, and the stress evolution is simulated by thermal–elastic–plastic FEM [7,11].

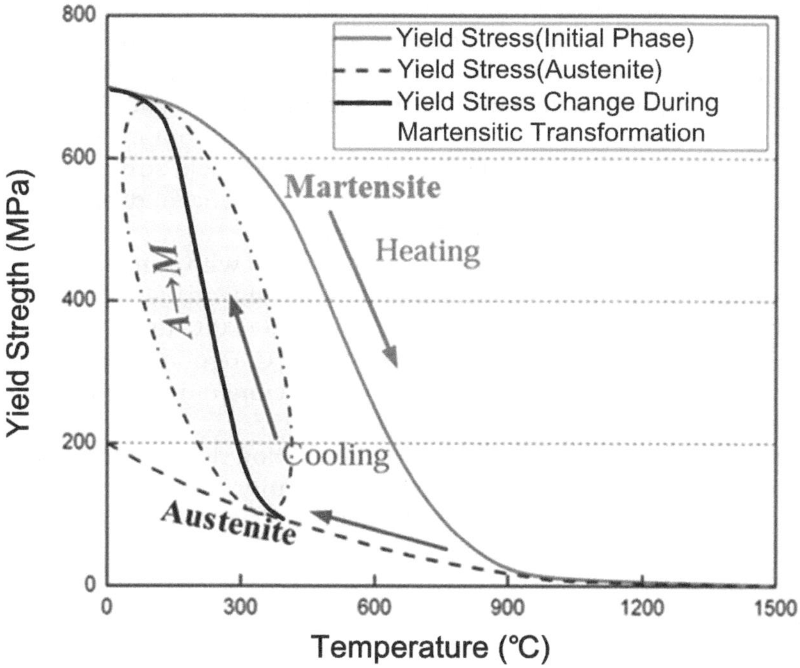

FIGURE 3.1.5
Yield strength change due to phase transformation during heating and cooling.

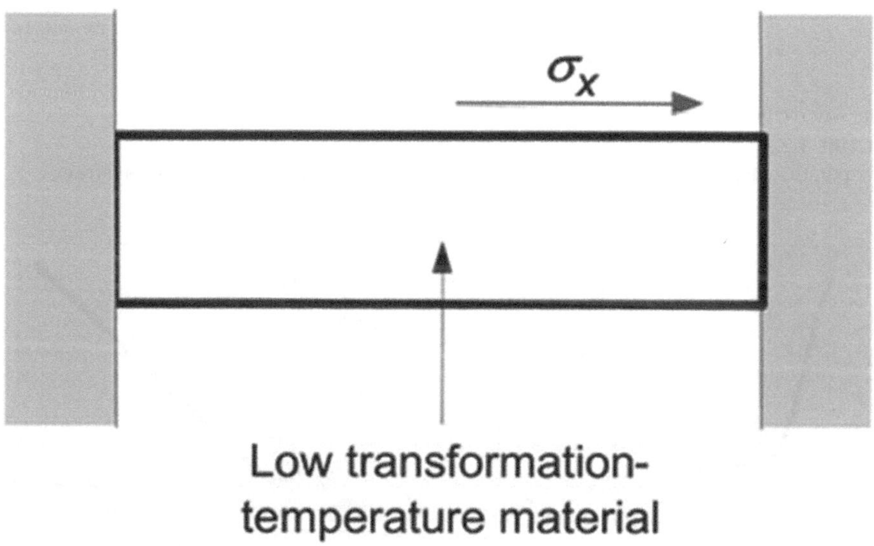

FIGURE 3.2.1
A simple bar model.

For comparison between an LTT material and a conventional (CONV) material, the yield strength and thermal strain for both materials and their temperature dependence are shown in Figures 3.2.2 and 3.2.3, respectively. The yield strength at room temperature is assumed to be 800 MPa.

The yield strength and thermal strain changes for both materials exhibit a similar tendency during the heating process. The yield strength decreases with increasing temperature and experiences a significant drop from 700°C to 800°C due to the austenite phase transformation.

The changes in yield strength and thermal strain with temperature during the cooling process for both materials are absolutely different. The significant change for the CONV material occurs from 600°C to 200°C. However, the change for the LTT material occurs from 240°C to room temperature, corresponding to the martensitic transformation temperature range. The thermal expansion for this LTT material is about 0.25%.

Figure 3.2.4 shows the simulated stress evolution during the cooling process for both CONV and LTT materials in a fully constrained bar model. The tensile stress in the CONV material increases significantly from 600°C to

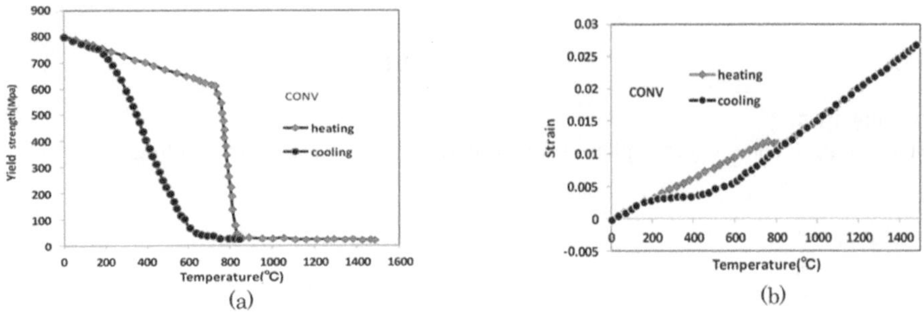

FIGURE 3.2.2
Temperature dependence of a CONV material: (a) yield strength and (b) thermal strain.

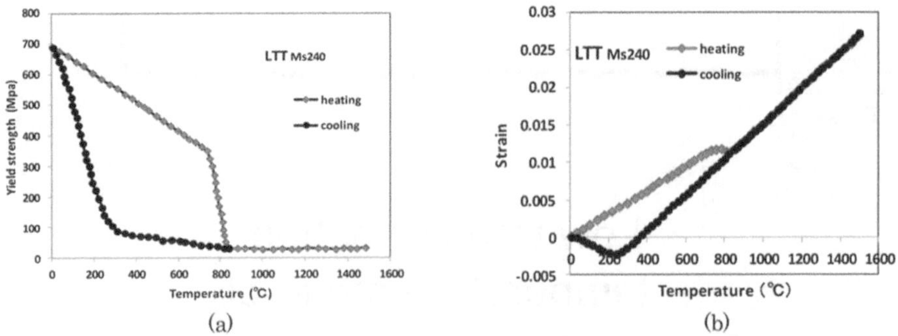

FIGURE 3.2.3
Temperature dependence of an LTT material: (a) yield strength and (b) thermal strain.

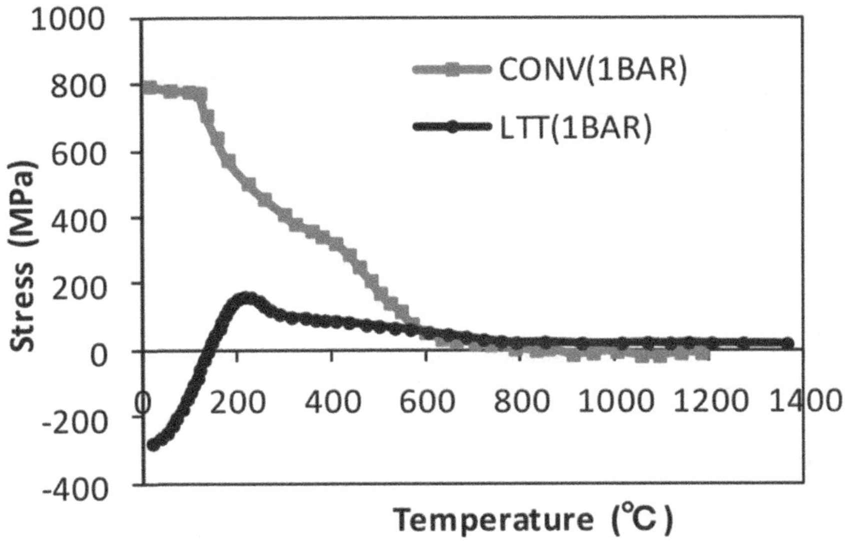

FIGURE 3.2.4
Thermal stress evolution in a fully constrained bar.

room temperature. The tensile thermal stress in the LTT material increases slightly from 600°C to Ms=240°C, following the curve of yield strength with temperature. When the temperature changes from Ms=240°C to about 150°C, the tensile thermal stress reduces to zero. With further cooling to room temperature, compressive residual stress is generated.

The compressive residual stress for this LTT material is about 300 MPa when Ms=240°C, and the volume expansion from Ms to room temperature is 0.25%. If the martensitic transformation expansion is larger, the compressive residual stress will increase.

Since the constraint in the welding direction is relatively strong and close to the fully constrained bar model, the stress change during cooling, shown in Figure 3.2.4, is similar to the stress component in the welding direction.

Figure 3.2.5a shows a bar model with two elements, where only one element is subjected to the heating and cooling thermal cycle. The temperature in the other element is maintained at room temperature. Their sizes are L1 and L2, respectively. The larger size of L1 means that the weld width is wide. The large size of L2 indicates that the constraint from the right fixed wall is far from the weld metal, resulting in a weak constraint on the weld metal.

As shown in Figure 3.2.5b, a large compressive residual stress in a fully constrained bar can be obtained. If L2 is larger and the ratio L1/L2 is smaller, the residual stress in the LTT WM may become tensile. This residual stress is similar to the residual stress component in the welding transverse direction.

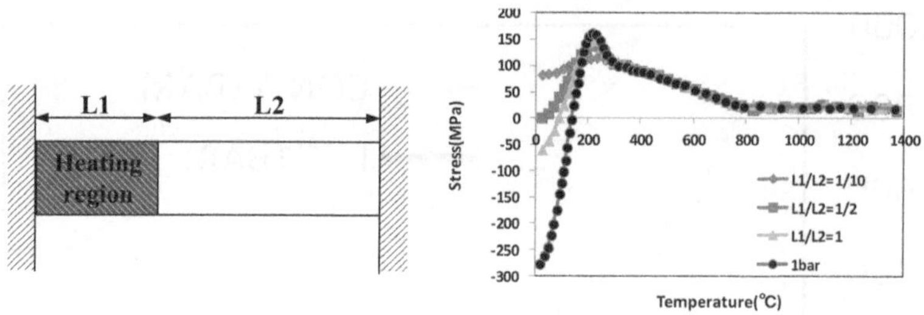

(a) A bar model with two elements. (b) Thermal stress evolution.

FIGURE 3.2.5
Thermal stress evolution in a constrained bar with different sizes of two elements. (a) A bar model with two elements. (b) Thermal stress evolution.

3.3 Residual Stress in Bead-on-Plate Joint

As shown in Figure 3.3.1, two bead-on-plate welded joints with different plate widths (80 and 200 mm) were fabricated using a semi-automatic gas metal arc welding process. The welds were deposited on 12-mm-thick SS400 steel plates using a commercial 3%Ni–3%Mn LTT covered electrode (Kobelco: LB-3AD). Table 3.3.1 lists the chemical composition and relatively higher martensitic transformation temperatures (Ms=398°C, Mf=230°C) of the LTT weld metal. The welding current, voltage, and speed were 130 A, 22 V, and 12.5 cm/min, respectively. Figure 3.3.1 also depicts the groove geometry before welding and the resulting molten zone after welding. Following welding, residual stress measurements were performed using the contour method.

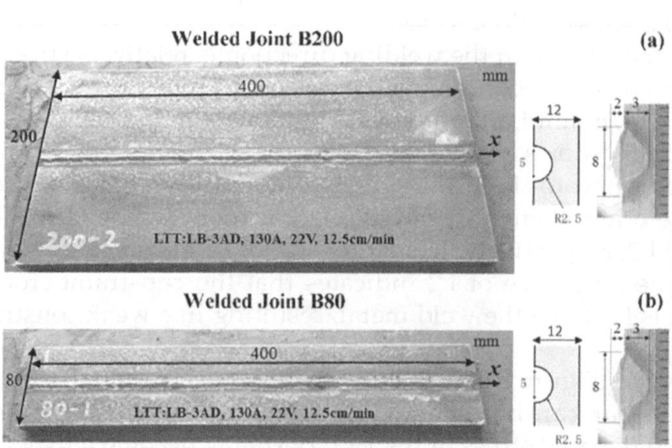

FIGURE 3.3.1
Two bead-on-plate welded joints: (a) B200 and (b) B80.

TABLE 3.3.1

Chemical Composition (wt.%) and Martensitic Transformation Temperatures (°C) of LTT Weld

C	Si	Mn	Ni	Ms	Mf
0.034	0.41	3.3	3.3	398	230

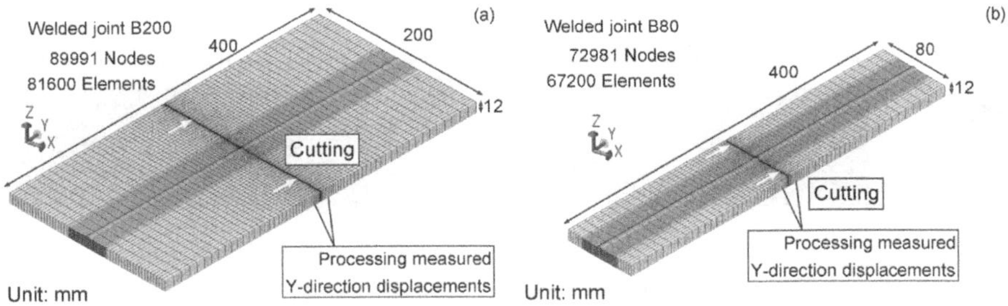

FIGURE 3.3.2
FE models applied to reproduce residual stress for the contour method.

As shown in Figure 3.3.2, the welded specimens were cut at the mid-transverse section perpendicular to the welding direction using wire electrical discharge machining. Subsequently, a three-dimensional elastic FEM was developed for residual stress calculation. The mesh was refined in the cutting zone, with element dimensions of 0.05 mm (welding direction), 1.0 mm (width direction), and 0.6 mm (thickness direction). In contrast, a coarser mesh (maximum element size: 4 mm) was employed in regions distant from the cutting zone. Material behaviors were set as isotropic and linearly elastic, with Young's modulus of 210 GPa and Poisson's ratio of 0.3.

3.3.1 Measured Residual Stress by Contour Method

Figure 3.3.3 presents the two-dimensional longitudinal residual stress distributions in welded joints B200 and B80 obtained through the contour method. The results demonstrate a significant reduction of tensile residual stresses in the LTT weld zone, counterbalanced by tensile stresses in the BM. Notably, compared to joint B200, joint B80 exhibits substantially higher compressive residual stresses.

For further quantitative comparison, Figure 3.3.4 plots the longitudinal residual stress (σ_y) distributions along the mid-section of the top surface. The LTT weld in joint B80 develops a pronounced compressive stress of approximately –200 MPa, while joint B200 shows relatively lower stress magnitudes ranging between –50 and +20 MPa in its LTT weld region.

The observed difference in residual stress mitigation between these two welded joints primarily stems from variations in plate width and associated

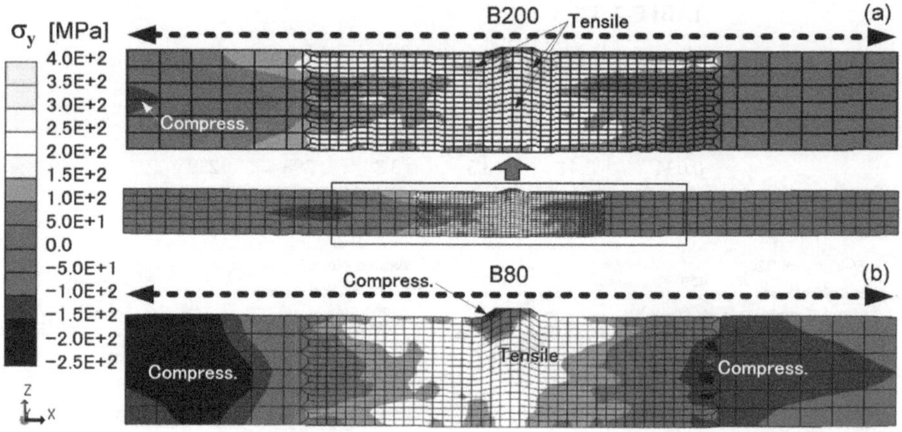

FIGURE 3.3.3
Distribution of residual stress σ_y on the cutting surface of welded joints.

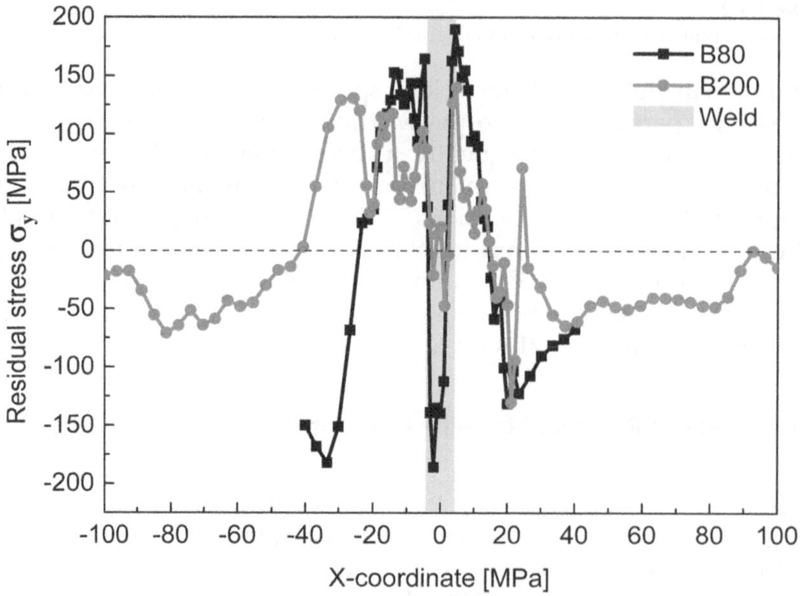

FIGURE 3.3.4
Distribution of measured residual stress σ_y on the top surface of welded joints.

heat dissipation characteristics. The wider B200 joint exhibits greater plate stiffness compared to the narrower B80 joint, as plate stiffness typically increases with width. Notably, the reduced plate width in joint B80 leads to lower heat dissipation during welding, significantly influencing the residual stress distribution. This observation aligns with established findings for conventional welds, where increased plate stiffness and enhanced heat dissipation typically lead to higher residual stresses in the weld zone [12,13].

To systematically investigate these two governing factors on residual stress evolution in LTT welded joints, a comprehensive thermal–elastic–plastic FE analysis was conducted, as detailed in the following section.

3.3.2 Analysis of Residual Stress in Two Experimental LTT Welded Joints

A thermal–elastic–plastic FEM was established to analyze the residual stress in two experimental LTT welded joints with different widths, prior to conducting parametric studies on the size effect. The simulation models, along with their respective boundary conditions, are illustrated in Figure 3.3.5, where the arrows indicate nodal displacement restraints. In the weld zone, the element dimensions were set to 5.0 mm along the welding direction, 0.5 mm in the width direction, and 0.5 mm in the thickness direction. In contrast, a relatively coarser mesh was adopted for regions remote from the weld zone, with a maximum element size of 5 mm. Welding parameters were defined in accordance with the experimental procedures.

Figure 3.3.6 presents the temperature-dependent thermophysical and mechanical properties of the LTT weld metal. Additionally, Figure 3.3.7 illustrates the transient thermal expansion coefficients of the LTT weld metal, as determined from dilatometric analysis.

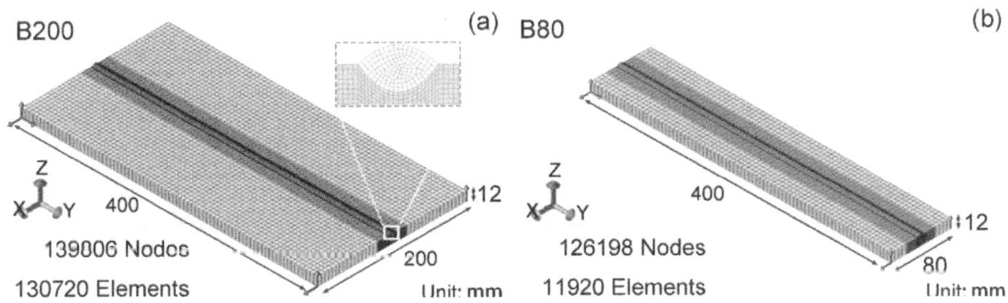

FIGURE 3.3.5
Simulation models and boundary conditions: (a) B200 and (b) B80.

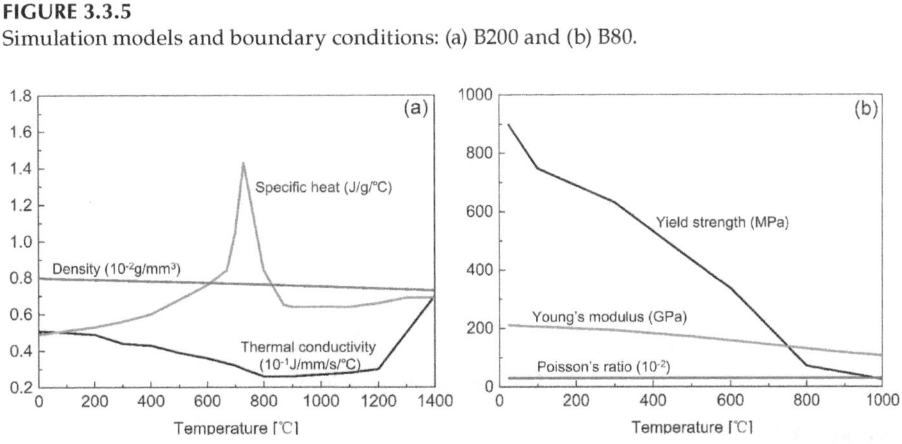

FIGURE 3.3.6
Material properties of LTT weld metal: (a) thermal properties and (b) mechanical properties.

As shown in Figure 3.3.8, the comparison of the longitudinal residual stress distributions on the top surface of the mid-section between the FEM and the contour method indicates good agreement.

To clarify the mechanism underlying residual stress development in welded joints B200 and B80, the thermal cycles and stress evolutions of selected nodes in the LTT weld and welded plate are plotted in Figure 3.3.9. As shown in Figure 3.3.9a, no significant difference in peak temperature is observed between the selected nodes of the two welded joints. Additionally, both the peak temperature and temperature gradient diminish at node N. Due to the size effect, the thermal cycles during the cooling process of welded joint B200 are notably steeper than those of B80. To assess the internal constraint of the welded joint, the isothermal temperature is defined as the point when the temperature difference between nodes M and N becomes zero.

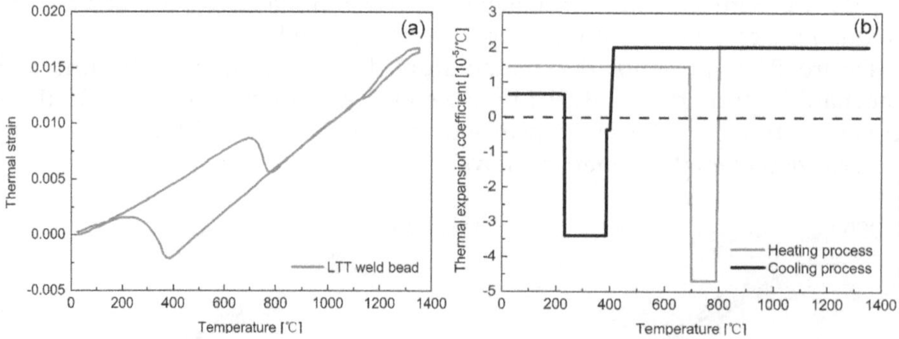

FIGURE 3.3.7
Dilatometric results of LTT weld: (a) dilatometric curve and (b) thermal expansion coefficients.

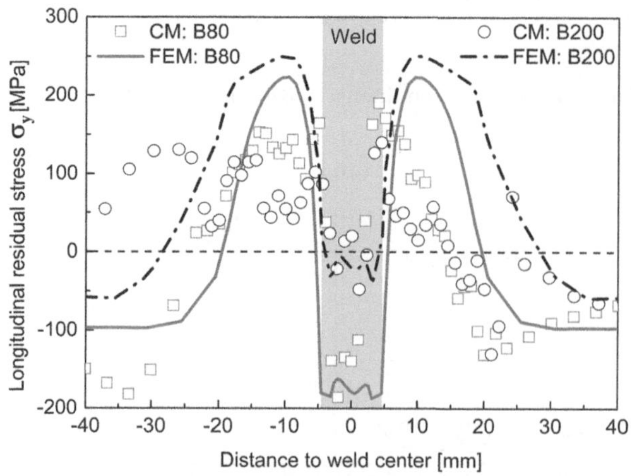

FIGURE 3.3.8
Computational results versus experimental measurements on the top joint surface.

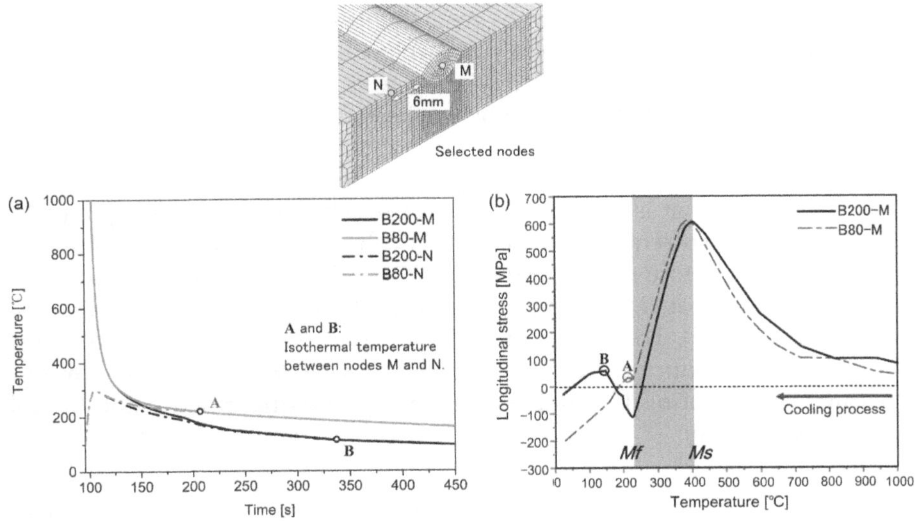

FIGURE 3.3.9
Thermal cycles and longitudinal stress evolution at selected nodes at the mid_section: (a) thermal cycles; (b) stress-temperature history during cooling.

Evidently, welded joint B200 requires a longer cooling time to reach the isothermal temperature between nodes M and N. The isothermal temperatures for B80 and B200 are approximately 210°C (point A) and 110°C (point B), respectively, with corresponding cooling times of about 210 and 338 seconds. Figure 3.3.9b presents the stress evolution at selected node M in the weld metal throughout the cooling process. During cooling, the tensile stress in both welded joints increases steadily prior to martensitic transformation. Upon martensitic transformation, the longitudinal stresses in both LTT welds decrease gradually. Owing to the high stiffness in wider plate joint B200, a compressive residual stress as low as –120 MPa is generated as the temperature falls to Mf temperature (230°C). In the subsequent cooling stage before reaching isothermal temperature B, the residual stress in the LTT weld of joint B200 shifts toward a tensile state due to thermal contraction and constraint from the base metal. Then, during the uniform cooling process from isothermal temperature B to room temperature, the stress in the LTT weld metal of joint B200 changes from tensile to compressive due to the difference thermal expansion coefficient between the base metal and the LTT weld metal. In narrow plate joint B80, the constraint on the LTT weld metal is relatively weak, and the tensile stress decrease resulting from martensitic transformation is smaller than that in joint B200. Moreover, because the isothermal temperature in joint B80 is higher (210°C) and close to the Mf temperature, the stress change over this small temperature range is very limited. Finally, compressive residual stress is generated in the LTT weld metal of joint B80 during uniform cooling from the isothermal temperature A (210°C) to room temperature, owing to the difference in thermal expansion coefficient, which is further discussed below.

Conventional understanding suggests that residual stresses should monotonically increase during cooling due to accumulated thermally induced strains. However, the observed deviation from this trend can be attributed to two key factors: (1) the difference in isothermal temperatures between the joints, and (2) the mismatch in thermal expansion coefficients between the LTT weld metal and the base plate. As demonstrated previously, joint B80 reaches a higher isothermal temperature (210°C) compared to joint B200 (110°C), while also requiring less time to achieve thermal equilibrium (210 vs 338 seconds). More significantly, the LTT weld metal exhibits a markedly lower thermal expansion coefficient ($6.5\times10^{-6}\ K^{-1}$ below 230°C) than the base plate ($1.5\times10^{-5}\ K^{-1}$). This substantial disparity in thermal expansion behavior leads to pronounced shrinkage adjacent to weld zone during continuous cooling. The resulting thermal contraction can be quantified using the following expression:

$$\Delta\varepsilon_{\text{th}} = \alpha\Delta T \tag{3.3.1}$$

where $\Delta\varepsilon_{\text{th}}$, α, and ΔT represent the increment of thermal strain, thermal expansion coefficient, and temperature interval, respectively.

Undoubtedly, the thermal shrinkage at node N on the welded plate is significantly greater than that at node M on the LTT weld. This indicates that the thermal contraction of the welded plate is constrained by the LTT weld metal during cooling, thereby increasing tensile stress in the BM while generating higher compressive stress in the LTT weld. Due to the differences in isothermal temperature and duration between the two welded joints, welded joint B80—which undergoes a higher isothermal temperature and shorter holding time—exhibits greater compressive residual stress.

3.3.3 Parametric Analysis of Residual Stress and Size Effect in 3%Ni–3%Mn LTT Welded Joints

In modern welding production, welded joints are fabricated with varying dimensions, resulting in differing residual stress distributions. To investigate the relationship between residual stress and LTT joint geometry, a series of numerical simulations were conducted on welded joints with varying widths and thicknesses while keeping other dimensions constant. Two parametric studies were performed: one examining the influence of plate width and the other analyzing plate thickness. Each study considered six distinct cases, with the corresponding dimensions and simulation parameters detailed in Tables 3.3.2 and 3.3.3, respectively. The welding parameters, material properties, and boundary conditions remained consistent with those outlined in the previous section.

TABLE 3.3.2

Simulation Cases Associated with Plate Width

Name	Plate Width	Length	Thickness	Number of Elements	Number of Nodes
B40	40	400	12	115,360	121,662
B60	60	400	12	117,280	123,930
B80	80	400	12	119,200	126,198
B140	140	400	12	124,960	133,002
B200	200	400	12	130,720	139,806
B400	400	400	12	149,920	162,486

TABLE 3.3.3

Simulation Cases Associated with Plate Thickness

Name	Thickness	Length	Plate Width	Number of Elements	Number of Nodes
T6	6	400	80	64,960	71,113
T8	8	400	80	85,440	92,178
T12	12	400	80	119,200	126,198
T20	20	400	80	139,040	146,610
T30	30	400	80	163,840	172,125

3.3.3.1 Effect of Plate Width on Residual Stress

Figure 3.3.10a and b presents the longitudinal residual stress distribution on the top surface of the mid-section, as well as the residual stress variation with plate width at the weld center and toe, respectively. The maximum compressive longitudinal residual stress (approximately −816 MPa) is generated in the LTT weld of joint B40. As the plate width increases to 200 mm, this stress increases significantly to about −23 MPa in joint B200. However, further increases in width (e.g., in joint B400) show a negligible effect on the longitudinal residual stress in the LTT weld.

In contrast, the longitudinal residual stress in the welded plate exhibits a slight increase with plate width. This trend can be attributed to the greater stiffness provided by the larger volume of cold BM in wider plates. While the tensile residual stress in the welded plate may approach the material's yield strength, failure typically initiates at critical stress concentration zones such as the weld toe and root. Notably, despite the high tensile residual stresses in the welded plate, extensive studies [14–16] have demonstrated that LTT weld materials significantly improve fatigue performance. Additionally, the weld toe consistently exhibits higher residual stresses than the weld center, with these stresses further increasing as plate width increases.

Figure 3.3.11 illustrates the correlation between residual stress and plate width at both the LTT weld center and toe. The longitudinal residual stresses at the weld center exhibit substantial variation within the 40–200 mm width

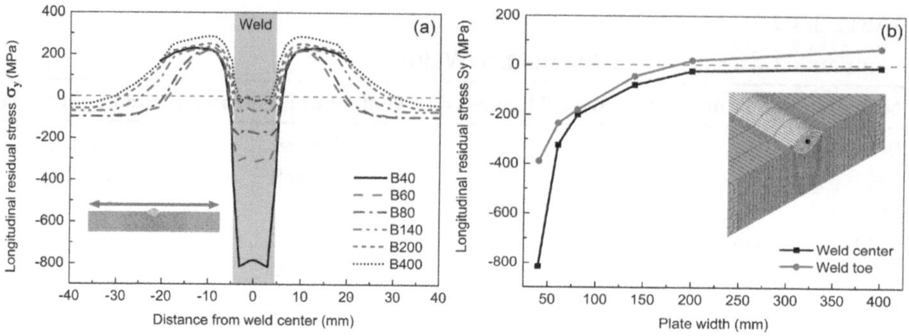

FIGURE 3.3.10
Computed longitudinal residual stress σ_y as a function of plate width (mid-section): (a) residual stress distribution on the top surface and (b) residual stresses at the weld center and toe.

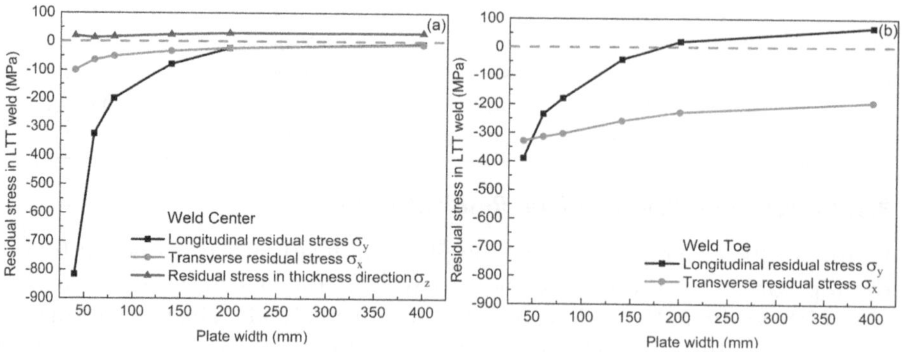

FIGURE 3.3.11
Residual stress as a function of plate width in 3%Ni–3%Mn LTT weld: (a) weld center and (b) weld toe.

range before stabilizing at plate widths exceeding 200 mm. A similar trend is observed for both longitudinal and transverse residual stresses at the weld toe. Notably, the longitudinal residual stress demonstrates a pronounced width-dependent increase, whereas other residual stress components are less sensitive to plate width variations.

3.3.3.2 Effect of Plate Thickness on Residual Stress

Figure 3.3.12 presents the relationship between residual stress and plate thickness in the weld zone. The residual stresses at the weld center undergo significant variations within the 6–20 mm thickness range. Specifically, the longitudinal residual stress shows a substantial increase between 6 and 12 mm thickness, followed by a gradual decrease as thickness approaches 30 mm. In contrast, the transverse residual stress exhibits a steady reduction with increasing thickness, while the through-thickness residual stress

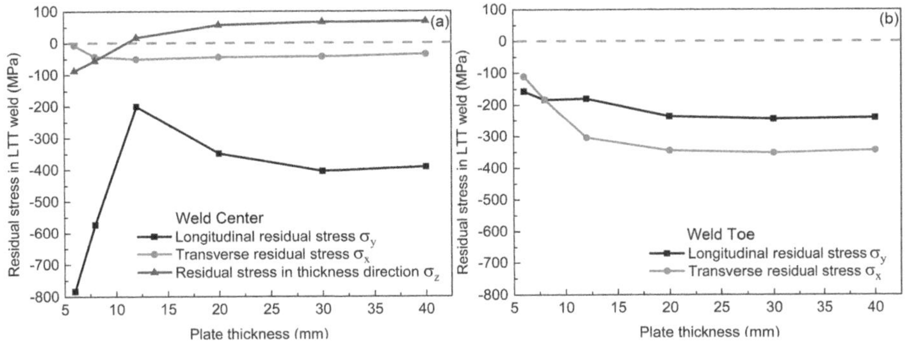

FIGURE 3.3.12
Residual stress as a function of plate thickness in 3%Ni–3%Mn LTT weld: (a) weld center and (b) weld toe.

demonstrates an inverse relationship. All residual stress components stabilize when the plate thickness reaches 30 mm. Evidently, compared to the plate width effect illustrated in Figure 3.3.11, thickness variation exerts a more pronounced influence on all residual stress components at the weld center. Regarding the weld toe region, the longitudinal and transverse residual stresses at the weld toe generally decrease with increasing plate thickness.

3.3.3.3 Effect of Ms Temperature on Residual Stress

As discussed in Section 3.3.1, the beneficial compressive longitudinal residual stress in LTT welds with an Ms temperature of 398°C diminishes with increasing plate width up to 200 mm. However, practical applications employ LTT welds with varying Ms temperatures. Notably, previous research [17] has demonstrated that LTT welds with an Ms temperature of 191°C generate the maximum compressive residual stress. This finding raises an important question regarding the potential influence of plate dimensions on the residual stress state in the LTT weld (Ms = 191°C), particularly since its Mf temperature falls below room temperature. Using the same simulation models and mechanical properties as in this study, residual stress calculations were performed to verify the size effect and the influence of Ms temperature.

Figure 3.3.13 illustrates the influence of Ms temperature on residual stresses at the weld center under different plate dimensions. Overall, the LTT weld with an Ms temperature of 191°C tends to generate higher compressive residual stresses. Furthermore, compared to those with an Ms temperature of 398°C, the residual stresses in LTT welds with an Ms temperature of 191°C are less sensitive to plate width. Conversely, when the plate width or thickness exceeds a certain threshold, further increases in plate dimensions have a negligible impact on residual stresses at the weld center. In summary, the LTT weld with an Ms temperature of 191°C is more conducive to the formation of compressive residual stresses at the weld center.

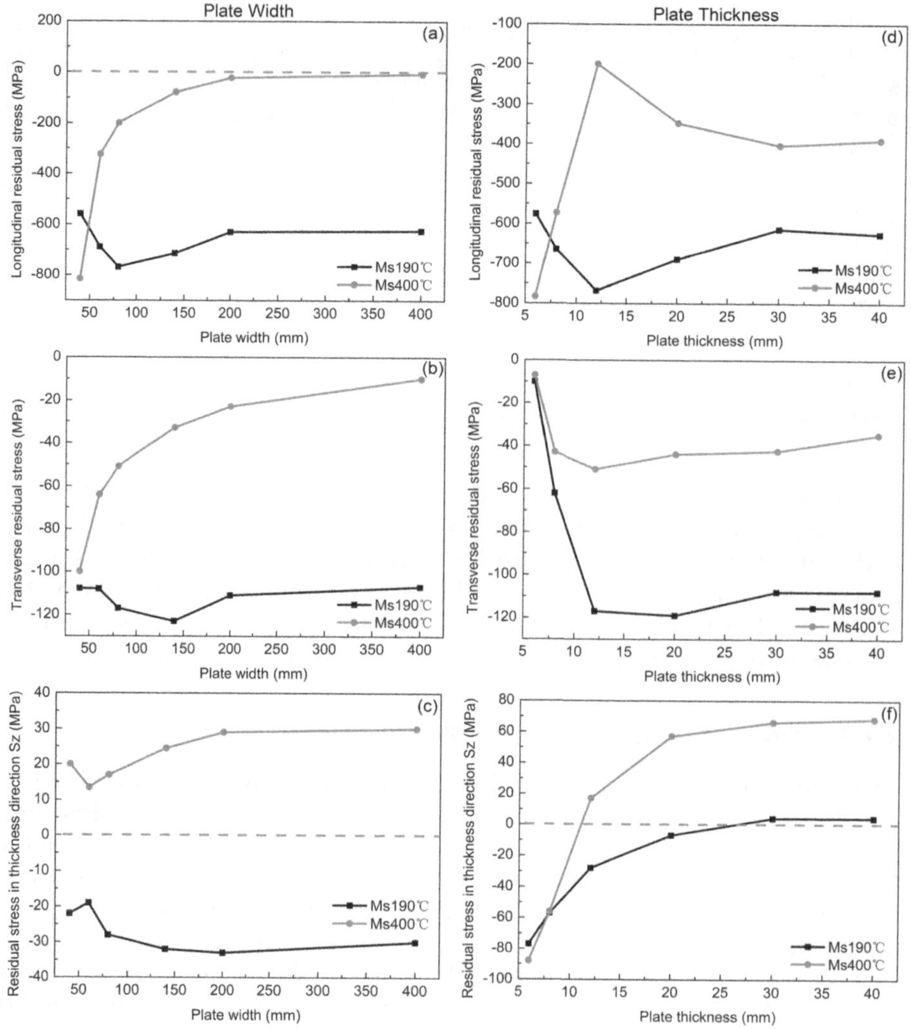

FIGURE 3.3.13
Comparison of residual stress at the weld center between two Ms temperatures under different plate dimensions: (a–c) effect of plate width and (d–f) effect of plate thickness.

Figure 3.3.14 demonstrates the effect of Ms temperature on residual stresses at the weld toe under varying plate dimensions. Overall, the LTT weld with an Ms temperature of 191°C results in higher compressive residual stresses. Furthermore, compared to those with an Ms temperature of 398°C, the residual stresses in the LTT weld with an Ms temperature of 191°C are less sensitive to plate width. Similarly, when the plate width or thickness exceeds a certain threshold, further increases in plate dimensions have a limited effect on altering residual stresses at the weld toe. Additionally, as shown in Figure 3.3.14d, the Ms temperature exerts only a minor influence on transverse residual stresses as the plate thickness changes. Overall, the LTT weld

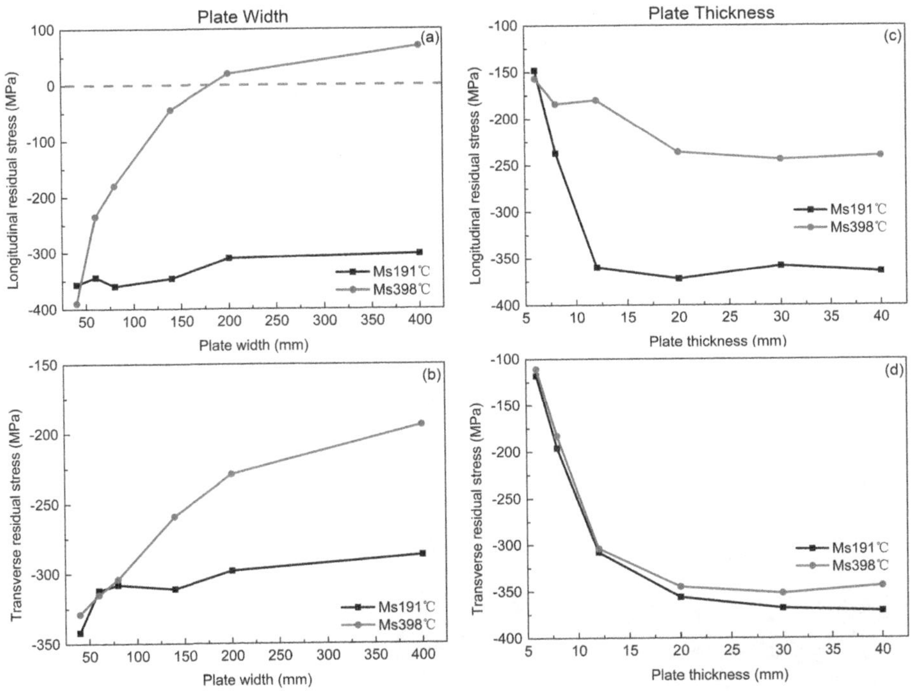

FIGURE 3.3.14
Comparison of residual stress at the weld toe between two Ms temperatures under different plate dimensions: (a and b) effect of plate width and (c and d) effect of plate thickness.

with the optimal Ms temperature of 191°C is conducive to generating higher compressive residual stresses at the weld toe.

3.4 Residual Stress in Butt-Welded Joints

High-strength steel plates (980 MPa tensile strength, 20 mm thick) were used to prepare two LTT butt-welded joints via MIG welding. The welding process involved a single pass per side in an X-groove, maintaining a heat input of 30–40 kJ/cm. Distinct solid weld wires, PS and PT, were employed for each joint. The PS wire was designed to yield a dual-phase martensite and austenite weld metal, possessing Ni and Cr equivalents of 9.2% and 12.8% (Ni_{eq} and Cr_{eq}), respectively. Conversely, the PT wire produced a fully martensitic weld metal, with Ni_{eq} and Cr_{eq} values of 7.0% and 12.0%, respectively.

Both PS and PT weld metals are characterized by remarkably low Ms temperatures compared to conventional weld metals. Their respective temperature-dependent dilatation curves are presented in Figure 3.4.1a and b. Specifically, the PS weld metal has Ms and Mf temperatures of 60°C and −170°C, while the PT weld metal's Ms and Mf temperatures are 244°C and 30°C.

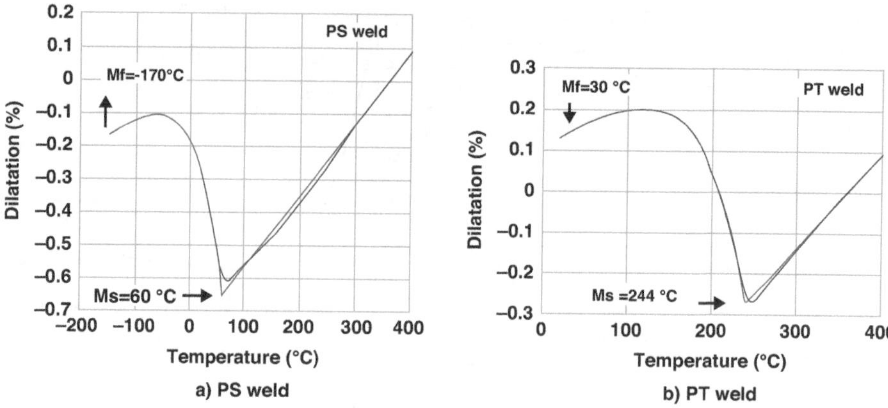

FIGURE 3.4.1
Temperature dependence of thermal dilatation of welds [18].

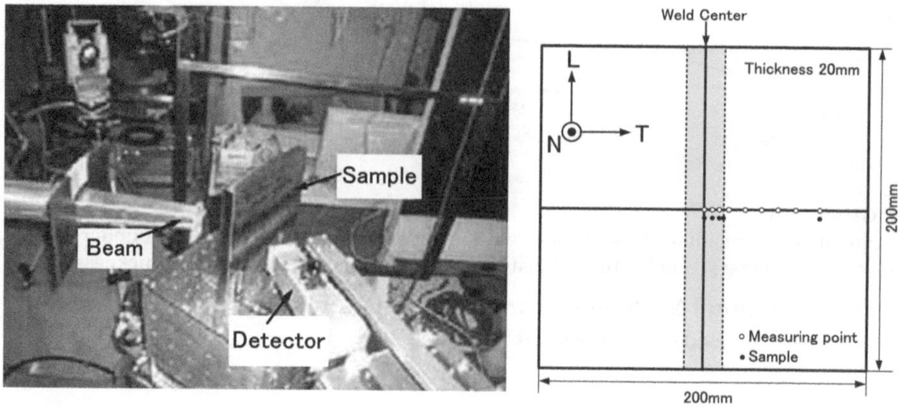

FIGURE 3.4.2
Residual stress measurement method: (a) experimental setup and (b) schematic drawing of welded joint indicating diffraction measurement positions [18].

Figure 3.4.2a and b presents an overview of the test piece during neutron diffraction measurement, along with a schematic diagram specifying the piece dimensions, including a weld length of 200 mm. The neutron diffraction measurements were conducted at a depth of 2 mm from the surface, with the beam position varied along the transverse direction, as indicated by the bold-marked "T" positions in Figure 3.4.2b. At each measurement point, the residual stress components in both the longitudinal (L) and transverse (T) directions of the welds were measured.

3.4.1 Microstructures of LTT Welds

Neutron diffraction peaks of stress-free standard specimens, cut from PS and PT weld metals, are illustrated in Figure 3.4.3a and b. An austenite peak

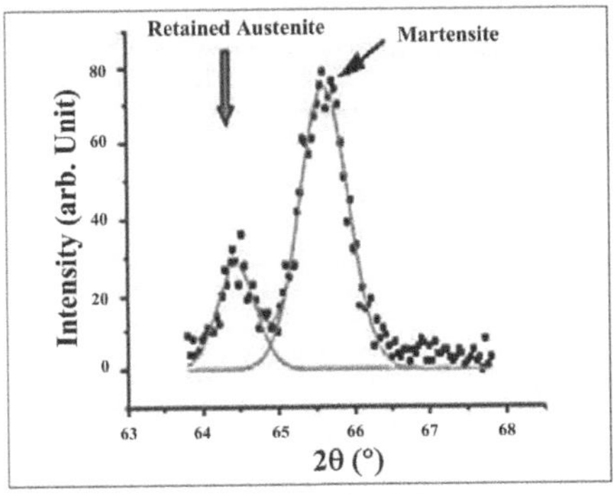

a) in PS weld metal

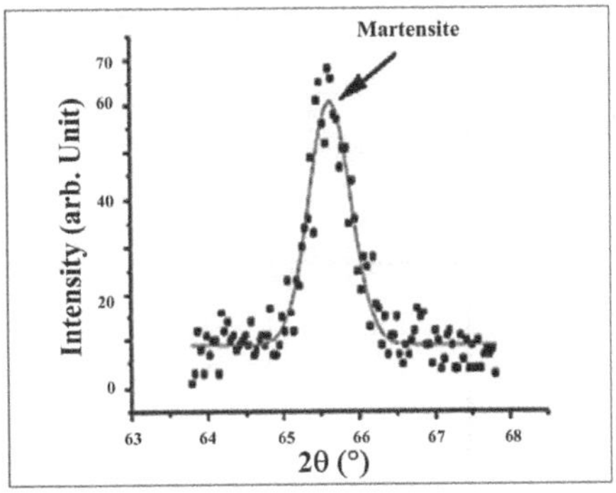

b) in PT weld metal

FIGURE 3.4.3
Neutron diffraction profiles of the microstructure in LTT metals [18].

was observed in the PS weld metal (Figure 3.4.3a), yet absent in the PT weld metal (Figure 3.4.3b). The volume fraction of the austenite phase, estimated from the intensities of peaks corresponding to austenite and martensite, is approximately 20%.

Based on these findings, the PS weld metal has a dual-phase microstructure containing roughly 20% austenite, with a Ms temperature of 60°C. Conversely, the PT weld metal features a fully martensitic microstructure, with an Ms temperature of 244°C.

3.4.2 Comparison of Residual Stress among Welded Joints

Figure 3.4.4 presents a comparison of residual stress distributions in welded joints, corresponding to the residual stresses in the longitudinal (L) and transverse (T) directions, respectively. For the LTT welded joints, the weld metal region spans ±13mm from the joint center, while the HAZ extends from ±13mm to approximately ±70mm. In both LTT weld metals, the residual stress within the weld metal is compressive in both L and T directions: specifically, a stress of –400 to –500MPa is observed near the center in the L direction, and approximately –100MPa in the T direction. The maximum tensile residual stress occurs in the HAZ adjacent to the weld metal, rather than at the weld toe position.

The residual stress difference between PS and PT weld metals is minimal; however, the compressive residual stress in the PS weld metal is slightly greater than that in the PT weld metal. In other words, the weld metal with an Ms temperature of 60°C and containing approximately 20% retained austenite exhibits a somewhat larger compressive residual stress compared to the fully martensitic weld metal with an Ms temperature of 244°C.

The longitudinal residual stress distribution differs significantly between LTT weld metals and conventional weld metals. Specifically, the LTT weld metals (60°C and 244°C) exhibit compressive residual stresses of approximately –400MPa within the weld zone, whereas the conventional weld metal shows tensile residual stresses reaching 450MPa.

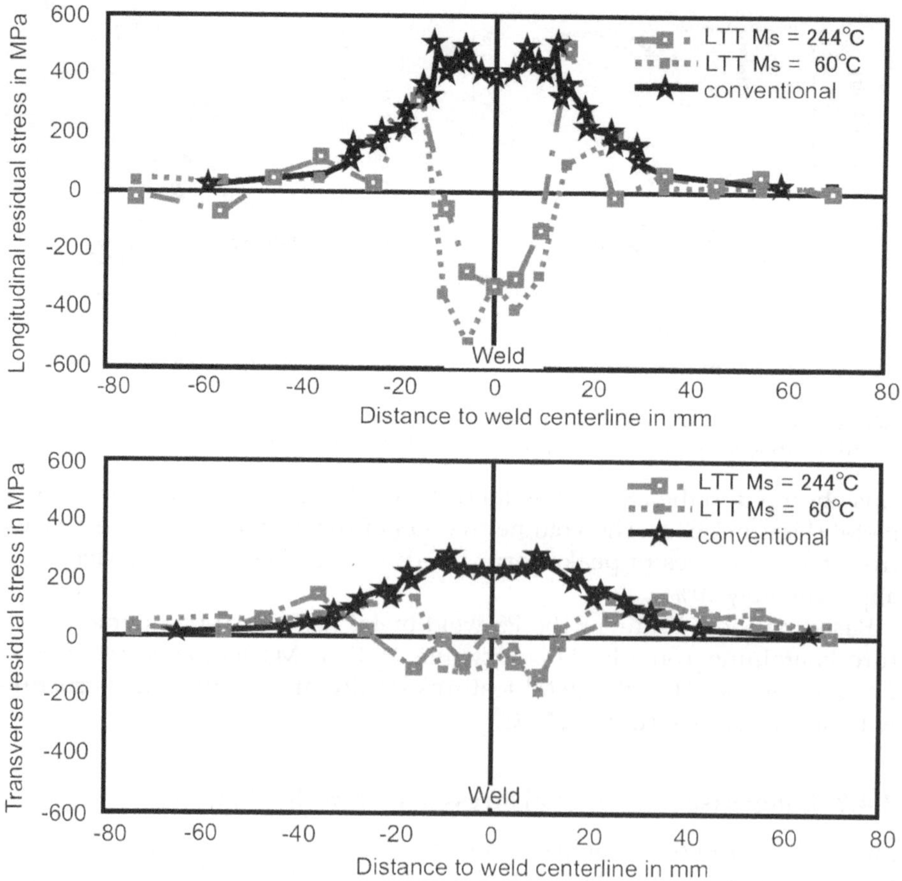

FIGURE 3.4.4
Longitudinal (above) and transverse residual stress (below) distributions of two LTT welded joints and a conventional high-strength joint [17,18].

Neutron diffraction measurements of residual stress clearly demonstrate that lowering the Ms temperature is highly effective in reducing residual stresses, even when the martensite volume fraction decreases from 100% to 80%. Furthermore, detailed residual stress distribution analysis reveals that the dilatation effect induced by martensitic transformation extends sufficiently to the weld toe, where residual stresses remain very low or even compressive.

As shown in Figure 3.4.3, the volume fraction of martensite is approximately 80% in the PS weld metal compared to 100% in the PT weld metal. Interestingly, despite exhibiting nearly identical amounts of thermal dilatation (Figure 3.4.1), the PS weld metal develops slightly higher compressive residual stresses within the weld region than the PT weld metal. This phenomenon can be explained by considering that the dilatation strain from martensitic transformation occurring in lower temperature regions generates greater compressive stress because Young's modulus of martensite increases as temperature decreases. Consequently, concurrent martensitic transformation at lower temperatures effectively produces high compressive stresses.

3.5 Residual Stress in Fillet-Welded Joints

3.5.1 Single-Side Pass Welded Joint

Transverse attachment joints fabricated via single-sided welding—similar to rib-to-deck connections in orthotropic steel bridge decks—were produced using LTT and conventional welding consumables, respectively. The configurations and dimensions of the specimens are illustrated in Figure 3.5.1. Table 3.5.1 presents the mechanical properties and chemical compositions of the steel used in the specimens. The main plate and rib plate were 12 and 6 mm in thickness, respectively, while the specimen width was 300 mm. The rib plate was joined to the main plate at a 77° angle, consistent with the typical configuration of rib-to-deck connections.

Single-side welding with a partial joint penetration groove weld was applied to the specimens. CO_2 gas-shielded arc welding with a 4.6%Mn LTT wire 1.2 mm in diameter and a conventional wire 1.4 mm in diameter (Kobelco: MX-Z200) was employed to fabricate the specimens. These wires are of the flux-cored type. Table 3.5.1 presents the mechanical properties

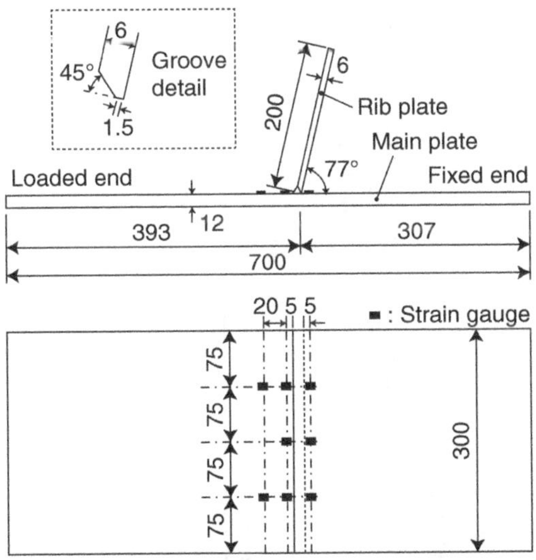

FIGURE 3.5.1
Schematic of the specimen [19].

TABLE 3.5.1

Mechanical Properties and Chemical Compositions of the Materials [19]

Materials		(a) Mechanical Properties			(b) Chemical Compositions (mass% Fe rem.)									
		Yield Strength (N/mm^2)	Tensile Strength (N/mm^2)	Elongation (%)	C	Si	Mn	P	S	Cu	Ni	Cr	Mo	V
Steel	Main plate	319	439	31	0.17	0.09	0.60	0.014	0.004	/	/	/	/	/
	Rid plate	289	441	27	0.16	0.02	0.52	0.016	0.006	/	/	/	/	/
Deposited Metal	LTT	760	811	20	0.03	0.42	4.6	0.008	0.007	0.01	0.01	0.02	0.01	0.01
	Conventional	478	572	26	0.03	0.47	1.53	0.009	0.008	0.01	0.01	0.02	<0.01	0.01

and chemical compositions of the deposited metals from these consumables. Welding parameters (Table 3.5.2) were controlled to achieve >75% penetration of the rib thickness (tr), meeting the requirement for rib-to-deck joints in the specifications for highway bridges in Japan [20]. As illustrated in Figure 3.5.2a, welding was performed with both the main (horizontal) and rib plates rigidly restrained at a fixed angle. Cross-sectional macrographs in Figure 3.5.2b and c confirm sound weld profiles for both LTT and conventional consumables, with no observed cracking in any specimen.

Residual stresses near the weld toe were measured using the X-ray diffraction method in both LTT and conventional weld specimens. The measurement direction was perpendicular to the weld. Figure 3.5.3 presents the transverse residual stress distribution, comparing experimental results (solid lines) with analytical predictions (broken lines). The plot shows residual stress (ordinate) versus transverse distance from the plate width center (abscissa). The LTT specimen has tensile residual stresses slightly smaller than those of the conventional specimen.

TABLE 3.5.2
Welding Conditions

Welding Consumable	Current (A)	Voltage (V)	Welding Speed (mm/min)	Heat Input (J/mm)	Preheating (°C)
LTT	250	28	370	1,135	50
Conventional	285	26.5	400	1,133	None

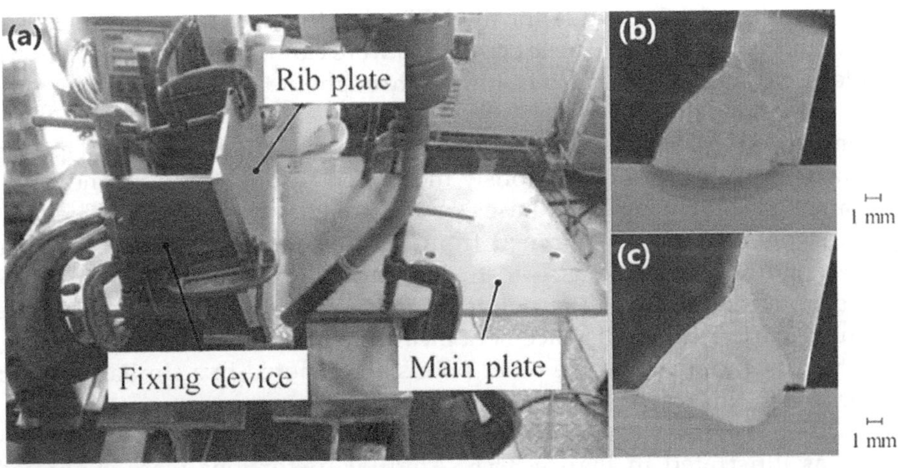

FIGURE 3.5.2
Specimen fabrication: (a) welding situation, (b) LTT, and (c) conventional bead [19].

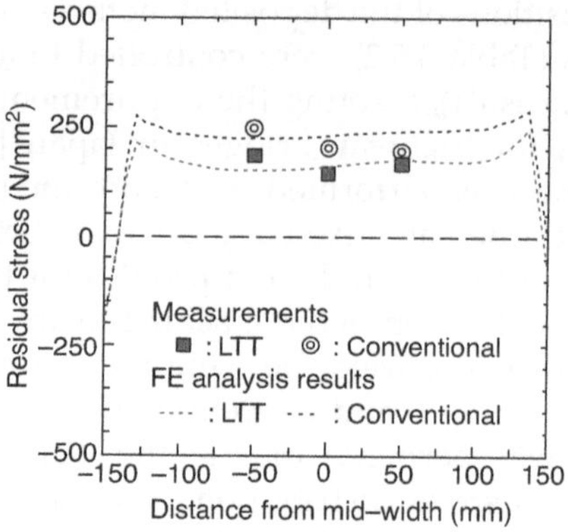

FIGURE 3.5.3
Residual stress distributions in the transverse direction [19].

3.5.2 Two-Side Single-Pass Welded Joint

The fillet T-joint depicted in Figure 3.5.4 was fabricated using a commercial steel for welded structures, SM490, with a thickness of 9 mm. Two different welding wires were employed: a conventional welding wire (Kobelco: MX-Z200) and a 10%Cr–10%Ni LTT wire. In contrast, the LTT welding wire features a Ni- and Cr-rich chemical composition, which effectively lowers the Ms temperature to Ms = 205°C. The chemical compositions of the materials are presented in Table 3.5.3.

The welding of both sides of the rib plate was carried out sequentially in opposite directions, utilizing 100% CO_2 arc welding. The interpass temperature between the first and second welding passes was maintained below 50°C. The welding parameters were set at 300 A, 30.5 V, and 450 mm/min for the conventional wire, while for the LTT wire, the parameters were 220 A, 29.5 V, and 400 mm/min. These conditions were carefully selected to achieve a fillet weld with a leg length of 6 mm. Throughout the entire welding process, the welded joints were left unrestrained.

Temperature profiles around the weld zone and displacement of points on the skin plate were continuously monitored throughout both the welding and subsequent cooling phases, as illustrated in Figure 3.5.5. The temperature measurements obtained from thermocouples T1 through T6 are correspondingly denoted as T_1 to T_6. The vertical displacements were measured from the top surface at points D1 to D6 and from the bottom surface at points D7 to D9. The displacement measured at point D1 is denoted as *d*1, and the displacements at the other points are denoted in a similar manner according to this convention.

Since the T-joint is unrestrained, angular distortion occurs asymmetrically, as illustrated in Figure 3.5.6. Angular distortions were evaluated at three locations along the welding direction, denoted as δ_1, δ_2, and δ_3, corresponding to their positions relative to the starting point of the first welding pass.

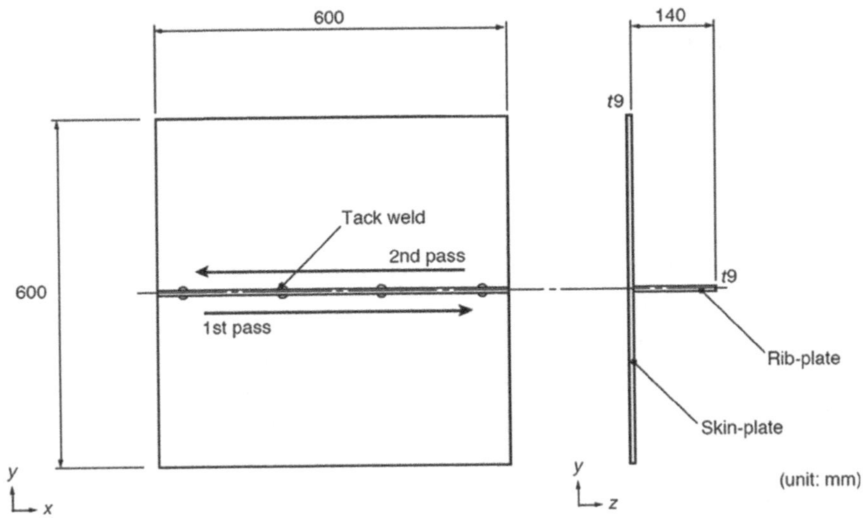

FIGURE 3.5.4
Schematic of T-joint [21].

TABLE 3.5.3
Chemical Compositions of the Welding Wires and the Steel Plate (wt-%)

Material	C	Si	Mn	P	S	Ni	Cr
LTT wire	0.020	0.39	0.19	0.010	0.006	10.14	9.76
Conventional Wire	0.044	0.51	1.59	0.013	0.008	0.01	0.020
Steel plate	0.14	0.28	1.44	0.015	0.004	/	/

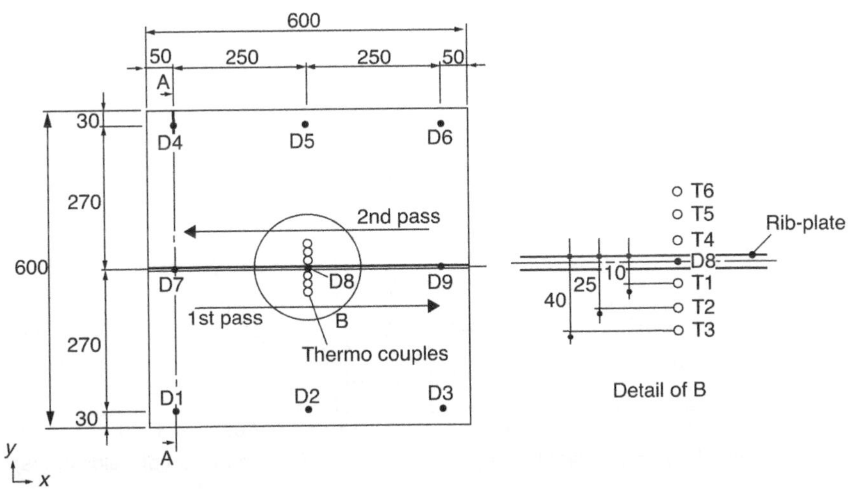

FIGURE 3.5.5
Location of measuring positions for displacement [21].

3.5.2.1 Angular Distortion of T-Joints between Conventional and LTT Welded Joints

Figure 3.5.7 presents the evolution of angular distortion under different welding conditions. When using conventional welding wire, the angular distortion initially increases before stabilizing. Conversely, with the LTT wire, the angular distortion reaches a peak value and subsequently decreases.

3.5.2.2 Numerical Simulation of Welding Distortion and Residual Stress of T-Joints

Three-dimensional FE analysis was employed to investigate the welding distortion and residual stress of the T-joints, incorporating the effect of phase transformation during welding into the analysis. The numerical simulation was carried out in two sequential steps:

First, the welding thermal cycles and microstructural evolution were calculated. A heat conduction analysis was performed to determine the welding thermal cycles. Concurrently, the microstructural evolution during these thermal cycles was computed using the continuous cooling transformation (CCT) diagrams of the materials involved.

Second, an elastic–plastic analysis was conducted to calculate the transient welding distortions and residual stresses. Based on the rule of mixtures, the

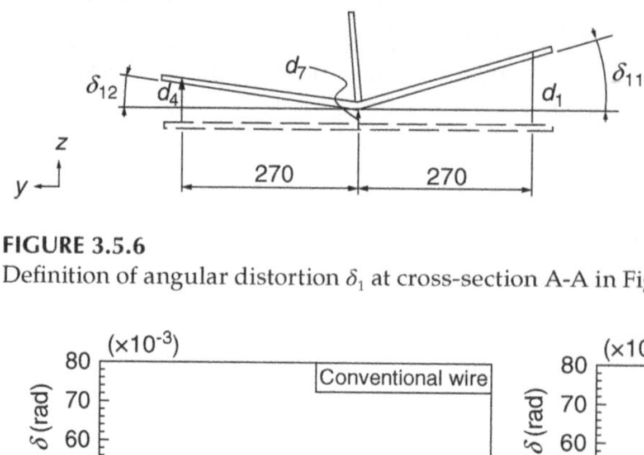

FIGURE 3.5.6
Definition of angular distortion δ_1 at cross-section A-A in Figure 3.5.5 [21].

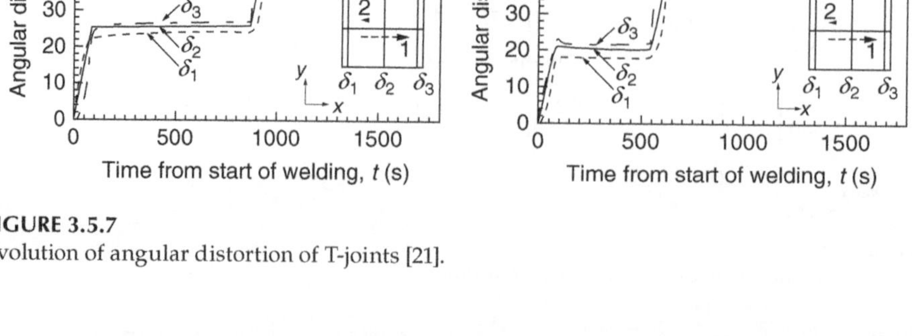

FIGURE 3.5.7
Evolution of angular distortion of T-joints [21].

mechanical properties of the materials at arbitrary temperatures during the welding thermal cycles were derived. These properties were determined from temperature- and microstructure-dependent mechanical parameters, combined with the temperature data and microstructure fractions obtained from the preceding heat conduction analysis.

3.5.2.2.1 Material Properties Used in the Numerical Simulation

The representative material properties utilized in the numerical simulation include CCT diagrams for calculating microstructural changes, as well as thermal expansion/contraction curves, Young's modulus, and yield stress for mechanical computations. These properties were determined through simulated welding thermal cycle tests and tensile tests at elevated temperatures. It should be noted that the measurements were conducted on specimens extracted from the weld metal of the T-joint, rather than on the welding wire itself. This is because the chemical composition—and consequently the mechanical properties—of the T-joint's weld metal differ from those of the original welding wire due to dilution from the steel plate. Table 3.5.4 presents the chemical compositions of the weld metals. A comparison between Tables 3.5.3 and 3.5.4 reveals that the nickel (Ni) and chromium (Cr) contents of the LTT welding wire decrease due to dilution from the steel plate.

The microstructural evolution during welding was predicted using the CCT diagrams presented in Figure 3.5.8. The temperature-dependent Young's modulus is illustrated in Figure 3.5.9, while Figure 3.5.10 shows the variation of yield stress as a function of both temperature and microstructure.

Figure 3.5.11 presents the thermal expansion/contraction characteristics of α and γ phases for each material (dashed lines), along with representative dilatometric curves obtained at a heating rate of 10°C/s and a cooling rate of 30°C/s (solid lines). For the LTT wire, the figure compares the transformation behavior between the original welding wire and the diluted weld metal. Notably, dilution from the steel plate elevates the Ms temperature (Ms) from 205°C to 380°C.

3.5.2.2.2 Welding Conditions in the Numerical Simulation

The parameters for the heat conduction analysis were determined by fitting the shape of the fusion zone. Figure 3.5.12 presents the isotherms of the welds computed from numerical simulations, together with the cross-sections of the T-joint welds.

TABLE 3.5.4

Chemical Compositions of Weld Metals

Material	C	Si	Mn	P	S	Ni	Cr
LTT weld metal	0.076	0.43	0.55	0.010	0.006	6.13	6.14
Conventional weld metal	0.10	0.45	1.62	0.013	0.006	0.01	0.01

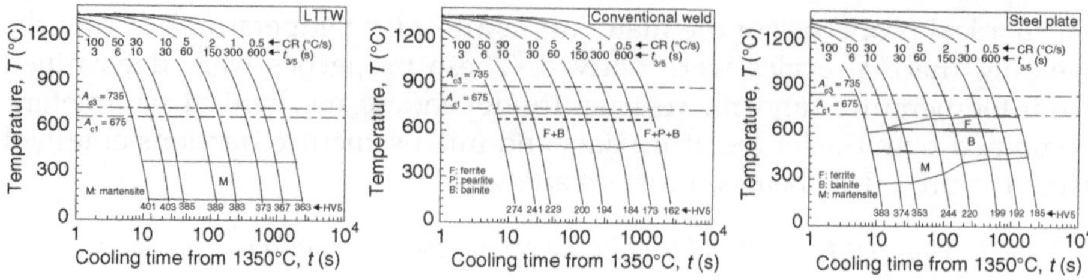

FIGURE 3.5.8
Continuous cooling transformation diagrams [21].

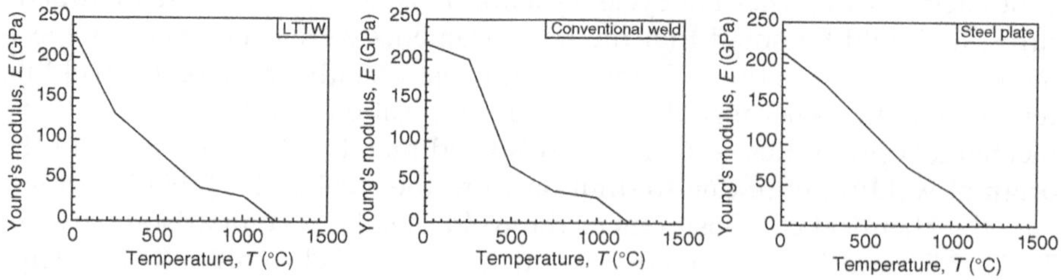

FIGURE 3.5.9
Temperature dependence of Young's modulus [21].

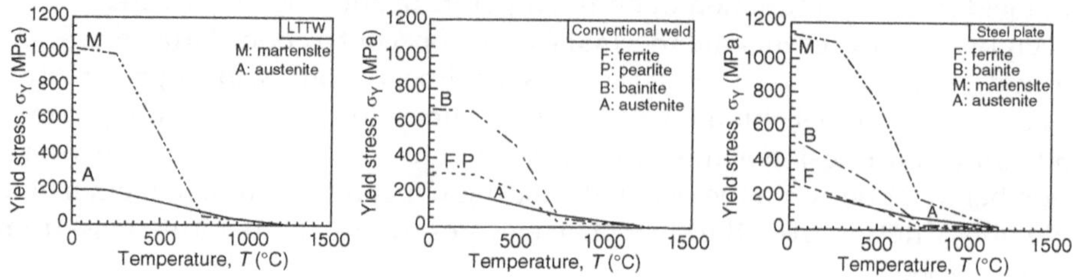

FIGURE 3.5.10
Temperature and microstructure dependence of yield stress [21].

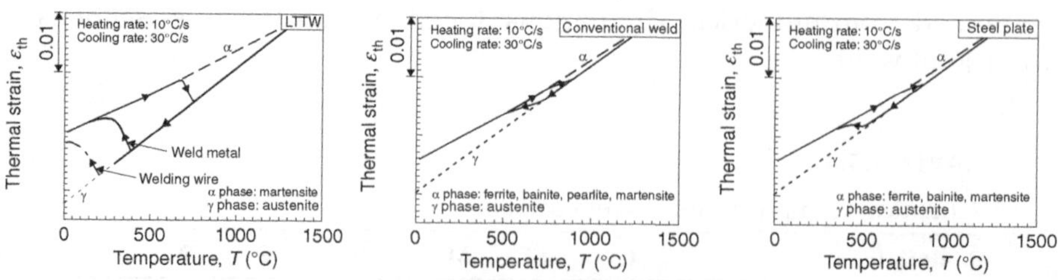

FIGURE 3.5.11
Temperature- and microstructure-dependent thermal strain [21].

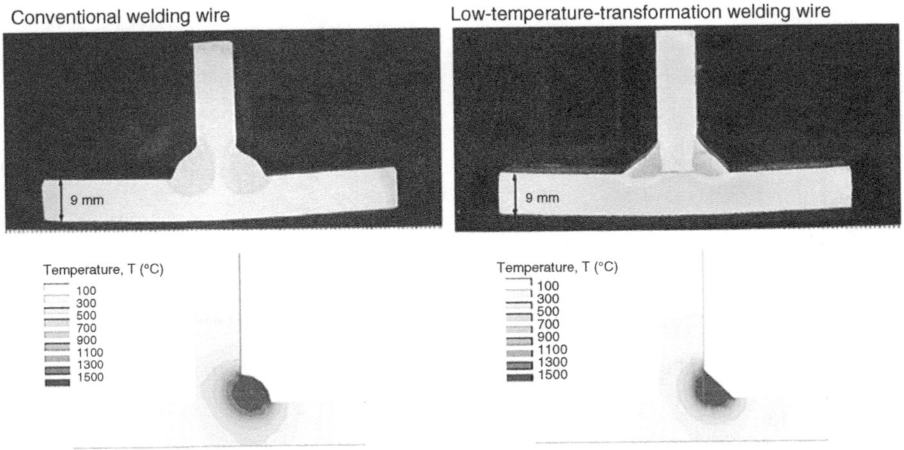

FIGURE 3.5.12
Cross-sections and isotherms of the welds [21].

3.5.2.2.3 *Simulated Results of Welding Distortion of T-Joints*

Figure 3.5.13 presents a comparative analysis of angular distortion (δ_2) between experimental measurements and numerical simulation results. The simulated angular distortions demonstrate good agreement with the experimental data.

Figure 3.5.14 presents an enlarged view (0–100 seconds) of the angular distortion evolution during the first welding pass for the T-joint using conventional welding wire. Angular distortion begins to increase with distance from the welding start position: δ_1 starts to rise first, followed by δ_2, and finally δ_3. The numerical simulations demonstrate high accuracy not only in calculating residual welding distortion but also in capturing the distortion behavior during welding.

Figure 3.5.15a presents the longitudinal bending distribution along the welding direction at the bottom surface of the skin plate. The results demonstrate that the LTT welding wire significantly mitigates this bending deformation. As shown in Figure 3.5.15b, the LTT wire provides an even more pronounced improvement in controlling longitudinal shrinkage, achieving substantial reduction compared to conventional welding wires.

3.5.2.2.4 *Evaluation of Welding Residual Stress*

Figure 3.5.16 presents the transverse distributions of the residual stress components (σ_x and σ_y). Consistent with previous research findings, the application of LTT wire demonstrates a significant reduction in residual stress magnitudes compared to conventional welding wire.

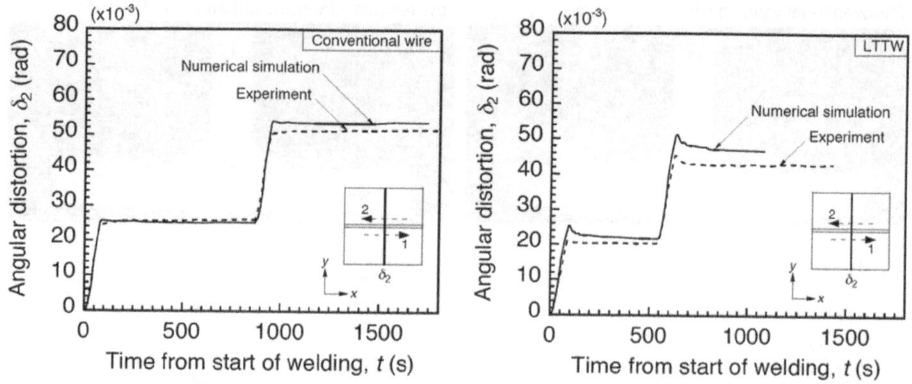

FIGURE 3.5.13
Comparison of angular distortion of T-joint between experiment and numerical simulation [21].

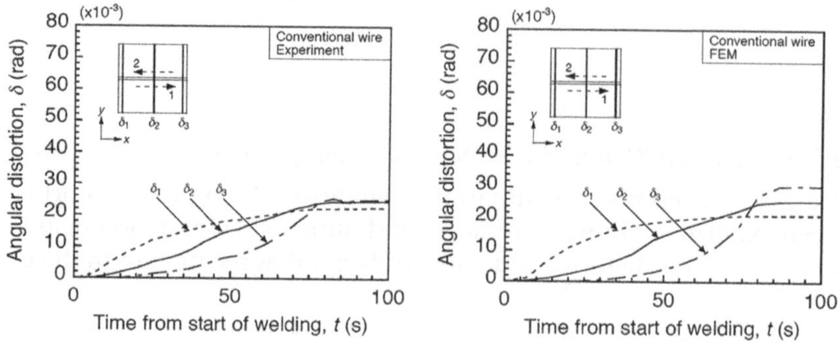

FIGURE 3.5.14
Enlarged view of welding distortion during the first welding pass [21].

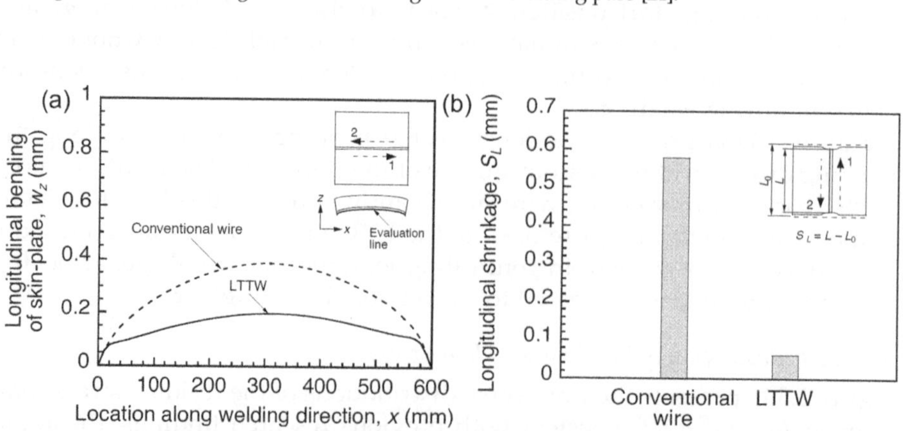

FIGURE 3.5.15
Comparison of longitudinal bending and shrinkage [21].

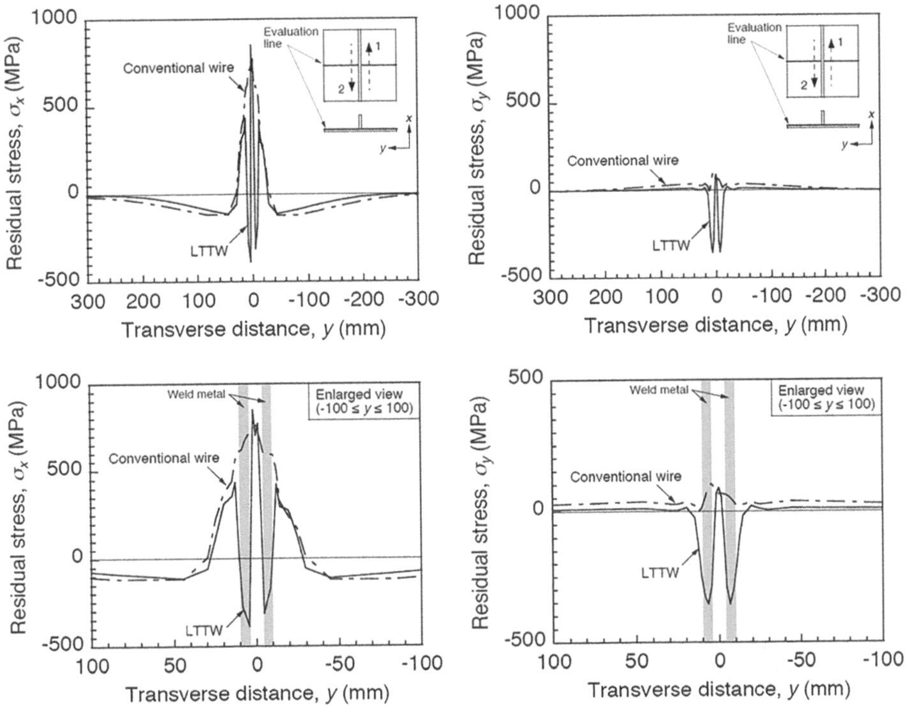

FIGURE 3.5.16
Distribution of computed residual stresses in the skin plate [21].

3.5.3 Multi-Pass Welded Joint

Unlike a single-pass weld, multi-pass welds undergo several thermal excursions as subsequent weld passes are deposited. Whether tensile residual stress can be induced in a multi-pass welded joint is greatly dependent on the reheat temperature. This section reveals the effect of the reheat temperature on the residual stress in multi-pass T-welded joints using an LTT welding rod. Accordingly, the residual stress distribution of the conventional T-welded joint is used for reference. FE analysis clarifies the residual stress development in the course during the welding process.

The schematic of a T-welded joint is displayed in Figure 3.5.17. Two T-welded joints were fabricated using 3%Ni–3%Mn LTT (Kobelco: LB-3AD) (Ms=398°C, Mf=230°C) and conventional (Kobelco: LB-47) welding electrodes. Accordingly, the former welded joint is classified as the LTT welded joint, while the latter is the conventional welded joint. The chemical compositions of the SM490A welded plates, the deposited metal, and the weld metal are listed in Table 3.5.5. During the welding process, the welding current, voltage, and speed were set at 130 A, 22 V, and 120 mm/min, respectively. Each weld pass was cooled to an ambient temperature of 15°C before the start of the subsequent weld pass.

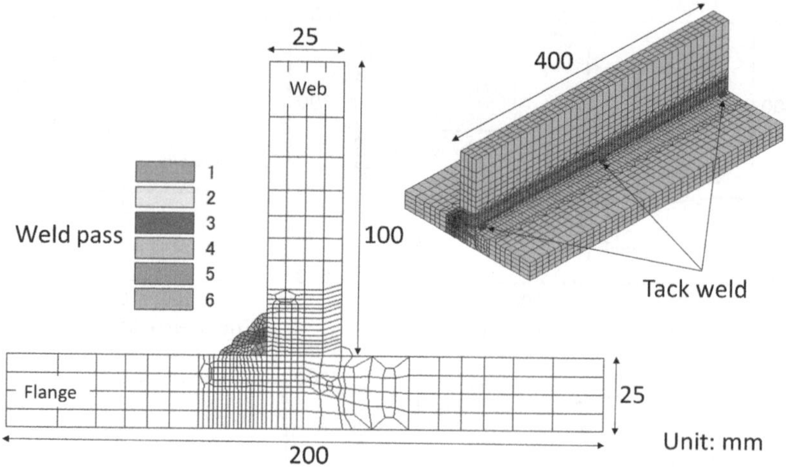

FIGURE 3.5.17
Schematic of a multi-pass T-welded joint.

TABLE 3.5.5
Chemical Compositions (%) of the Welded Plate, Deposited Metals, and Weld Metals

Name	Condition	C	Si	Mn	Ni
SM490		0.200	0.55	1.60	/
LB-3AD	Deposited metal	0.034	0.41	3.30	3.30
	Weld metal	0.067	0.34	2.95	2.65
LB-47	Deposited metal	0.080	0.55	0.79	/
	Weld metal	0.093	0.49	0.64	/

After fabricating the welded joints, residual stress measurement was conducted using the cosα method [22]. The transverse stress (vertical with respect to the welding direction) in the flange plate was measured.

3.5.3.1 Residual Stress Measurement Results

To compare the residual stress between the conventional and LTT welded joints, as shown in Figure 3.5.18, the residual σ_y stress was measured using the X-ray diffraction technique at three positions on the flange plate. The measured residual σ_y stress is plotted on the right side of Figure 3.5.18, which also shows the five moving averages intended to minimize experimental error. Evidently, the residual stress distribution is very similar among these three positions, demonstrating the high accuracy of the experimental measurement. Furthermore, the conventional welded joint has a lower tensile residual stress of about 100 MPa at the weld toe compared to approximately 250 MPa in the LTT welded joint. As the distance from the weld toe increases,

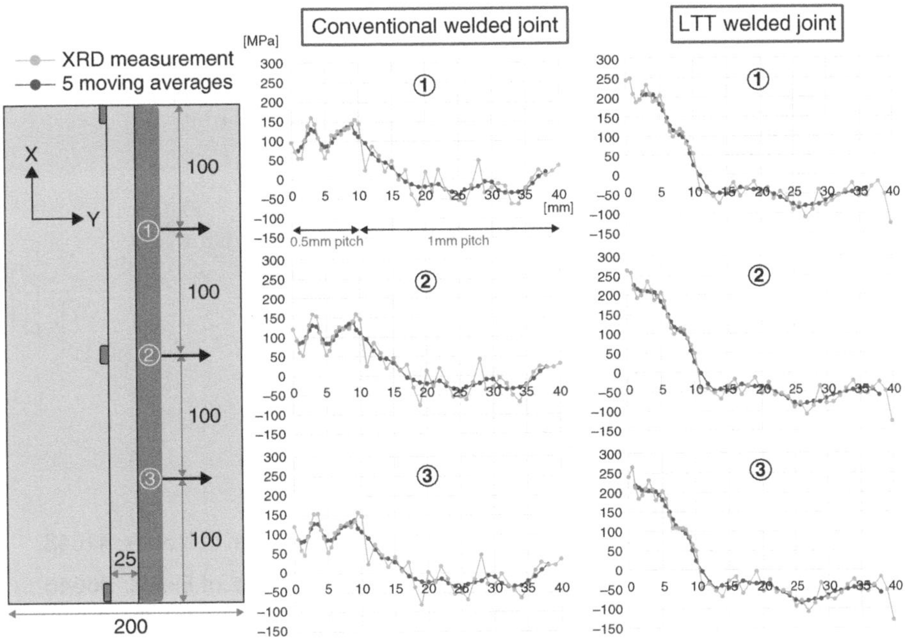

FIGURE 3.5.18
Measured residual σ_y stress of the two welded joints.

the tensile residual stress in the LTT welded joint gradually decreases. Accordingly, the reheat temperature is responsible for the higher tensile residual stress induced in the LTT welded joint.

3.5.3.2 FEM Simulation

To clarify the difference in the residual stress distribution between these two welded joints, a thermal–elastic–plastic analysis was implemented. Figure 3.5.19 shows the simulation model for computating the residual stress. The total numbers of nodes and elements are 41,648 and 36,646, respectively. A finer mesh of about 2 mm was adopted around the weld pass, while coarser meshes were utilized in regions away from the weld zone, with a maximum size of 8 mm. The boundary condition was defined by the arrows, which denote the fixed displacement in the corresponding direction. Six weld passes were deposited on one side of the T-welded joints, and each weld pass was cooled to an ambient temperature of 15°C before the start of the next weld pass. The welding parameters of the welding simulation were the same as in the experiment. In addition, the conventional von Mises elastoplastic flow theory was taken into account.

The thermal and mechanical properties of the LTT weld metals can be referred to in Figures 3.3.6 and 3.3.7. Except for the mechanical properties

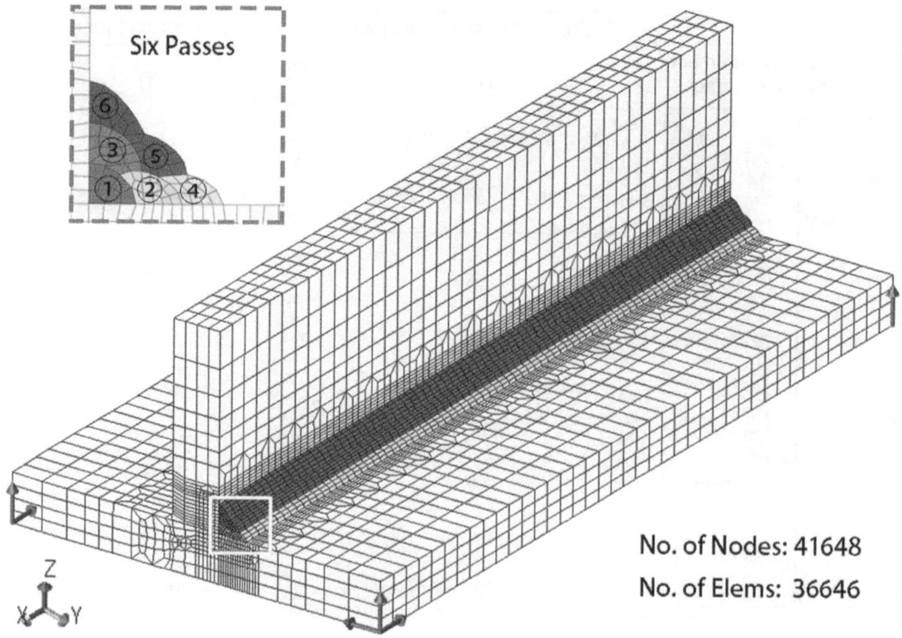

FIGURE 3.5.19
Simulation model for the computation of residual stress.

at room temperature, the other temperature-dependent mechanical properties of conventional (Kobelco: LB-47) and 3%Ni–3%Mn LTT weld metals were computed using thermodynamic software, JMatPro. The mechanical properties of SM490A were experimentally measured based on the specifications for rolled steels used in welded structures. In addition, Figure 3.5.20 displays the transient coefficients of thermal expansion (CTE) of the welded components, as measured by the dilatometric test. The CTE is given by [23]:

$$CTE = d\varepsilon / \Delta T \tag{3.5.1}$$

$$d\varepsilon = \Delta L / L_0 \tag{3.5.2}$$

where L_0 is the sample length at room temperature and ΔL is the temperature-dependent length change.

3.5.3.3 Reheat Temperature of LTT Weld Passes

In multi-pass welding, the previously formed weld beads undergo different thermal cycles. The reheat temperature plays an indispensable role in the residual stress development in an LTT welded joint. Hence, it is critical to evaluate the reheat temperature distribution induced by these thermal

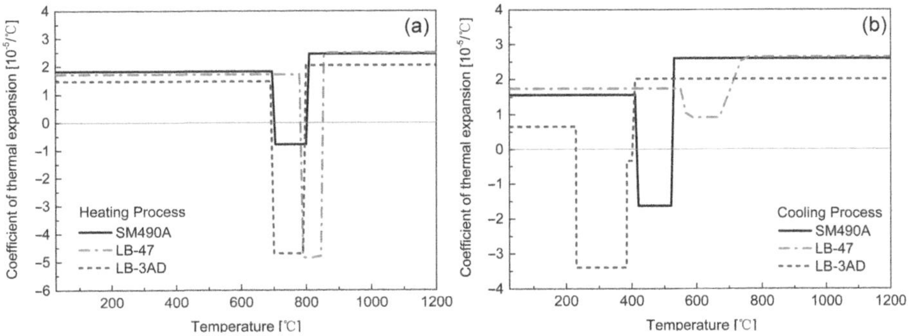

FIGURE 3.5.20
CTE of the welded components during the heating and cooling processes: (a) heating process and (b) cooling process.

cycles in the previously formed weld beads. Additionally, the reheat temperature has a strong relationship with the Ac3 temperature. The Ac3 temperature of LTT welds during the heating process is about 780°C. In regions with a reheat temperature below 780°C, martensite cannot fully transform into austenite, resulting in an increase in thermally induced tensile residual stress. In other words, the CTE of this region during the cooling process is the same as that during the heating process. If the reheat temperature exceeds 780°C, martensite will fully transform into austenite during the heating process, and then austenite transforms back into martensite during the cooling process, thus producing beneficial compressive residual stress. Figure 3.5.21 shows the temperature distribution of the weld passes after each welding. According to the temperature distribution and Ac3 tempe rature, regions with temperatures higher than 780°C can be defined as the "real" LTT weld zone, which can produce beneficial compressive residual stress after each welding. Conversely, regions with temperatures below 780°C are defined as the "non-real" LTT weld zone. Correspondingly, the CTEs of the "non-real" and "real" LTT weld metals are redefined again for the residual stress computation.

3.5.3.4 Comparison of the Residual Stresses between the Welded Joints

Figure 3.5.22 shows the residual σ_y stress distribution in the mid-section of the conventional and LTT welded joints. In the conventional welded joint, most of the weld zone experiences a greater tensile stress. For the LTT welded joint, compressive residual σ_y stress is mainly produced in the newly formed LTT weld zone with a reheat temperature higher than the Ac3 temperature, balanced by tensile stress in the nearby region. Due to the welding sequence, the residual stress state of the weld passes formed first converts from compressive to tensile because the reheat temperature is lower than the Ac3 temperature. Hence, greater tensile residual stress is also induced

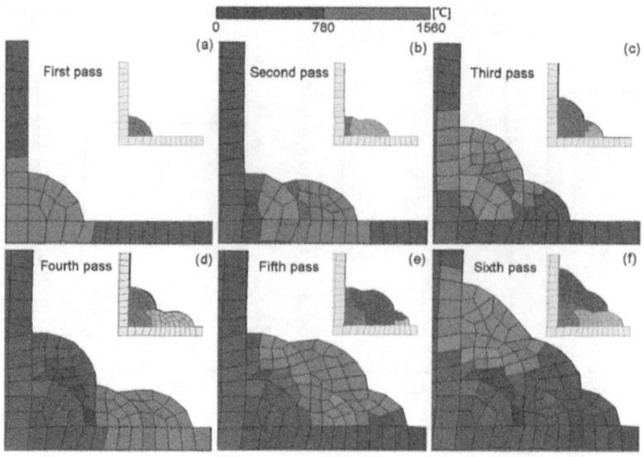

FIGURE 3.5.21
Temperature distribution during the welding process for the redefinition of the CTE of the LTT bead: (a) first pass, (b) second pass, (c) third pass, (d) fourth pass, (e) fifth pass, and (f) sixth pass.

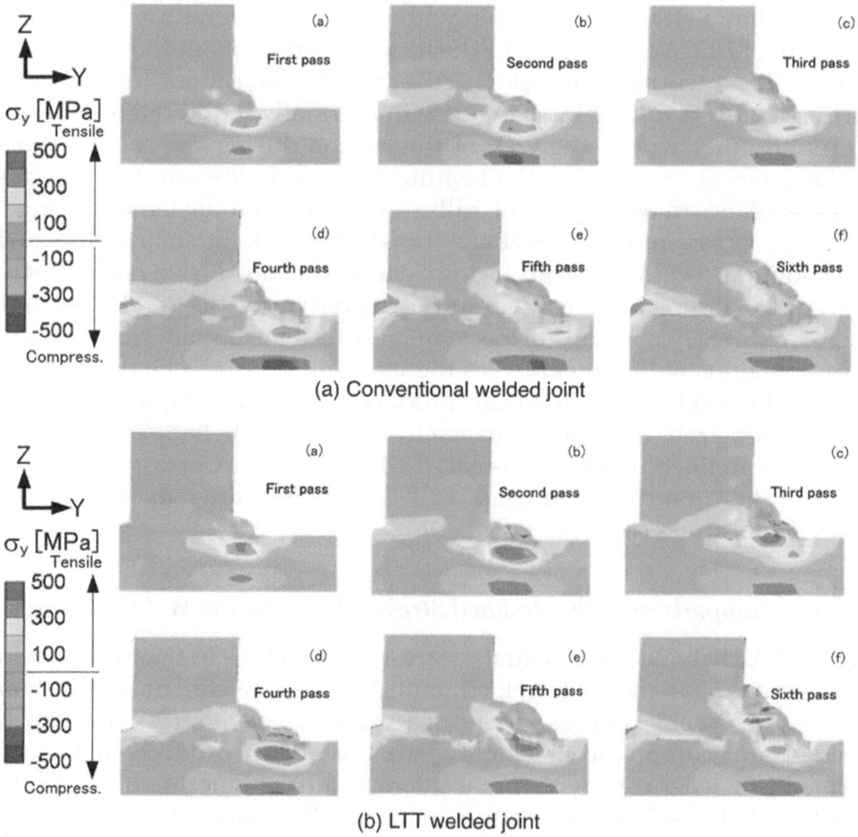

FIGURE 3.5.22
Residual stress σ_y distribution in the mid-section of the conventional and LTT welded joints:(a) first pass, (b) second pass, (c) third pass, (d) fourth pass, (e) fifth pass, and (f) sixth pass.

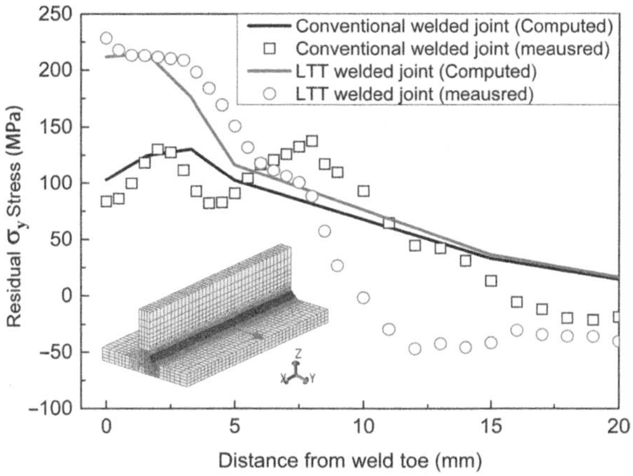

FIGURE 3.5.23
Computed residual σ_y stress distribution on base metal versus the experimental result.

at the weld toe of weld pass 4, while compressive stress is produced at the weld toe of weld pass 6. Moreover, the residual stress near the tack weld in both welded joints builds up gradually during the welding process, and the conventional welded joint also exhibits greater tensile stress.

Figure 3.5.23 compares the computed residual σ_y stress distribution on surface of base metal with the measured stress. It can be seen that the computed residual stresses align well with the experimental measurements adjacent to the weld toe.

3.6 Residual Stress in Boxing Fillet-Welded Joints

In this section, the effects of weld metal material and welding pass sequence on residual stress in the boxing fillet-welded joint are investigated both numerically and experimentally. Figure 3.6.1 displays the geometry and dimensions of a boxing fillet-welded joint, representative of an actual structural component. The assembly consists of a stiffener (150 mm × 50 mm × 19 mm) and a base plate (800 mm × 240 mm × 19 mm), both made of 780 MPa class high-strength steel, joined by single-pass fillet welds. For comparison, two types of weld metals were studied: a CONV material and a 10%Cr-10%Ni LTT material (Ms=205°C). The mechanical properties of the conventional weld metal were matched to those of the BM. Figure 3.6.2 presents the FEM used for residual stress simulation, with a heat input of 2.9 kJ/mm.

Regarding the welding sequences for the boxing fillet-welded joint, two types shown in Figure 3.6.3 were considered. In Sequence A, the first pass starts welding near the center of the stiffener's longitudinal direction,

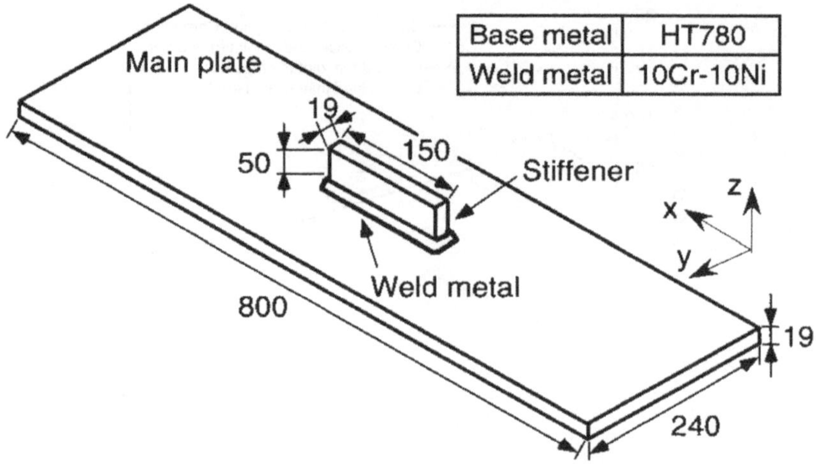

FIGURE 3.6.1
Configuration of a boxing fillet-welded joint between plate and stiffener [24].

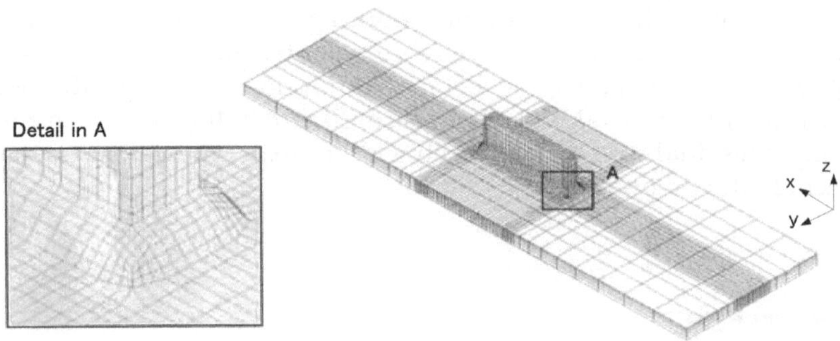

FIGURE 3.6.2
Finite element model used for residual stress simulation [24].

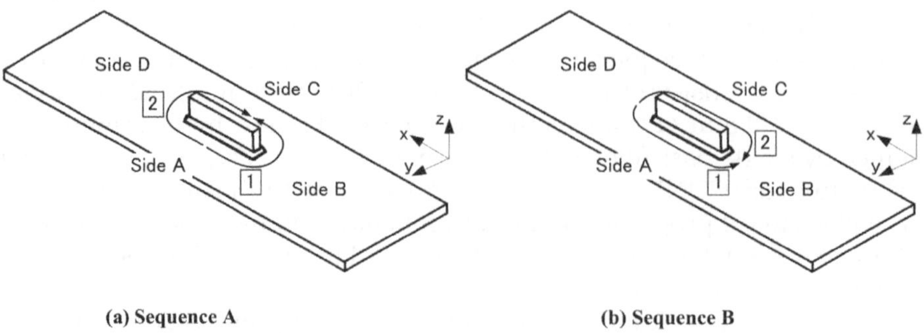

FIGURE 3.6.3
Welding sequences of a boxing fillet-welded joint between plate and stiffener [24].

continues around one end of the stiffener, and then proceeds along the opposite longitudinal side until reaching a point precisely corresponding to the back side of the welding start point. Subsequently, the second pass is welded on the opposite side of the stiffener. In Sequence B, the weld start and end points are located at the stiffener ends, and welding is completed in two passes without performing continuous all-around fillet welding. For both welding sequences, the interpass temperature was set to 50°C; the second pass began only after the entire specimen had cooled to the specified temperature following the completion of the first pass. Furthermore, based on the positional relationship between the base plate and the stiffener, the longitudinal sides of the stiffener are referred to as Side A and Side C, while the stiffener end sides are referred to as Side B and Side D, as shown in the figure. These designations are used to organize the subsequent analysis results.

Actual welded joints were fabricated for experimental validation under the same conditions as those used in the analysis, employing both LTT and conventional welding materials. The residual stress distribution near the weld ends was then measured using a strain gauge.

3.6.1 Comparison of Numerical Results and Measured Residual Stress Distributions

Figure 3.6.4 presents the residual stress analysis results for the case employing welding sequence A with 10%Cr-10%Ni LTT welding material. The longitudinal stress on the base plate is illustrated along two paths: (1) a transverse line crossing the plate width through the HAZ near the stiffener end, and (2) a longitudinal line along the base plate centerline, extending from the weld root at the stiffener end to beyond the weld toe. The results indicate that

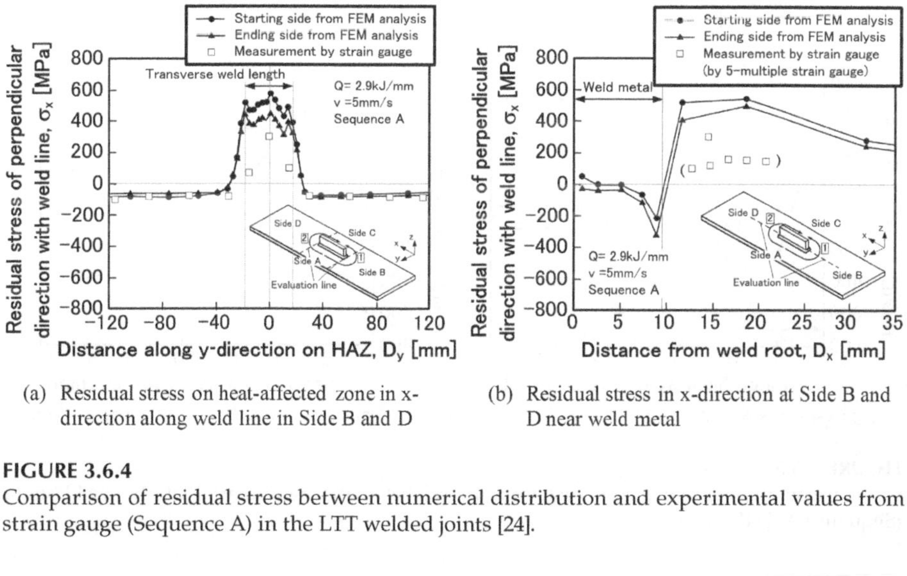

(a) Residual stress on heat-affected zone in x-direction along weld line in Side B and D

(b) Residual stress in x-direction at Side B and D near weld metal

FIGURE 3.6.4
Comparison of residual stress between numerical distribution and experimental values from strain gauge (Sequence A) in the LTT welded joints [24].

while tensile stress persists near the HAZ, the weld metal itself exhibits compressive stress throughout its entire area, or where tensile stress exists, the residual stress values are very small. Figure 3.6.4 also compares the numerical residual stress results with values measured on actual welded joints using strain gauges. The numerical results show generally good agreement with the experimentally measured residual stresses on the joint surface.

3.6.2 Influence of Welding Material and Welding Sequence on Residual Stress

3.6.2.1 Relationship between Welding Material and Residual Stress

Figure 3.6.5 shows a comparison of residual stress distributions for Sequence A when using LTT welding material versus conventional welding material for 780 MPa class steel. The longitudinal and transverse stresses on the base plate surface are displayed for (1) an extension line on the base plate surface at its center, running from the weld root at the stiffener center to beyond the weld end (Side A), and (2) an extension line along the longitudinal centerline of the base plate surface, running from the weld root at the stiffener end to beyond the weld end (Side B). It was confirmed that the residual stress distributions are nearly similar at geometrically symmetric positions relative to the stiffener (Side A and Side C, Side B and Side D). Although Side A and Side C correspond to the overlapping areas of the weld start and end points, respectively, they also show nearly similar residual stress distributions.

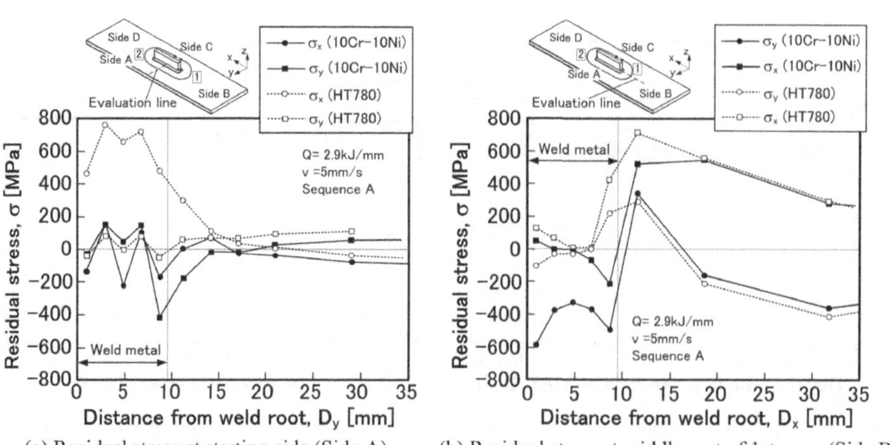

(a) Residual stress at starting side (Side A) (b) Residual stress at middle part of 1st pass (Side B)

FIGURE 3.6.5
Comparison of residual stress distribution by the changing material of the weld metal (Sequence A) [24].

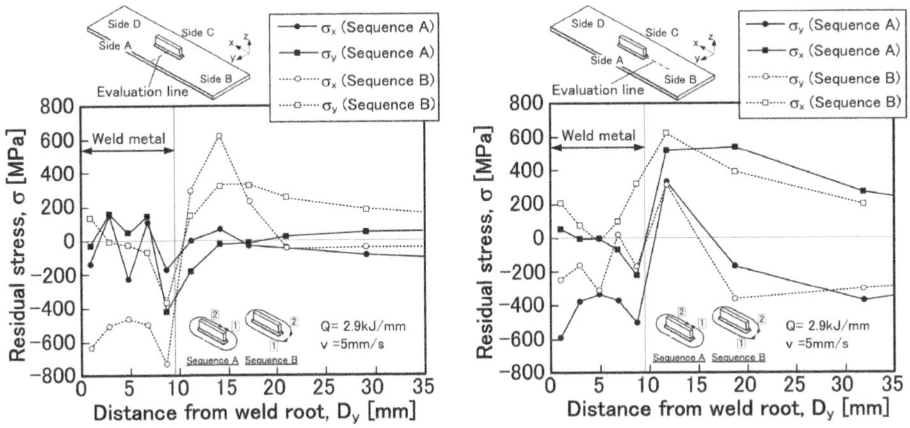

(a) Residual stress at Side A

(b) Residual stress at Side B

FIGURE 3.6.6
Comparison of residual stress distribution on the main plate and weld metal along the center lines of the stiffener by the changing sequence of welding deposition [24].

3.6.2.2 Relationship between Welding Sequence and Residual Stress

Figure 3.6.6 shows a comparison of residual stresses when the welding sequence of the passes is changed. Similar to Figure 3.6.5, the longitudinal and transverse stresses on the base plate surface are displayed for (1) Side A and (2) Side B. Although some tensile stress exists in parts of the weld metal, the residual stress is predominantly compressive regardless of the welding sequence. In the case of Sequence B, tensile residual stress occurs at the weld end; however, this value is confirmed to be significantly smaller compared to when using conventional welding material. That is, it is important to avoid overlapping the start and end points of weld beads at locations critical for strength design, not only from the perspective of weld defects but also from the perspective of residual stress. In any case, it can be concluded that using LTT welding material reduces the residual stress in the weld metal, regardless of the welding sequence.

References

1. N. Ma, et al., *Welding Deformation and Residual Stress Prevention*. Butterworth-Heinemann, Oxford (2022).
2. J. Goldak, *A Double Ellipsoid finite Element Model for Welding Heat Sources*. IIW Doc. No., 212 (1985).
3. Z. Feng, et al., Investigation of the residual stress in a multi-pass T-welded joint using low transformation temperature welding wire. *Materials*, 14(2) (2021), 325.

4. T. Inoue, Mechanism of transformation plasticity and the unified constitutive equation for transformation-thermo-mechanical plasticity with some applications. *International Journal of Microstructure and Materials Properties*, 5(4–5) (2010), 319–327.
5. T. Inoue, Mechanics and characteristics of transformation plasticity and metallo-thermo-mechanical process simulation. *Procedia Engineering*, 10 (2011), 3793–3798.
6. J.-B. Leblond, J. Devaux, and J. Devaux, Mathematical modelling of transformation plasticity in steels I: Case of ideal-plastic phases. *International Journal of Plasticity*, 5(6) (1989), 551–572.
7. J. Yamamoto, et al., Analysis of martensite transformation behavior in welded joint of low transformation-temperature materials. *Quarterly Journal of the Japan Welding Society*, 25(4) (2007), 560–568 (in Japanese).
8. S. Li, et al., Numerical investigation of formation mechanism of welding residual stress in P92 steel multi-pass joints. *Journal of Materials Processing Technology*, 244 (2017), 240–252.
9. D.P. Koistinen and R.E. Marburger, A general equation prescribing the extent of the austenite-martensite transformation in pure iron-carbon alloys and plain carbon steels. *Acta Metallurgica*, 7(1) (1959), 59–60.
10. D. Deng and H. Murakawa, Influence of transformation induced plasticity on simulated results of welding residual stress in low temperature transformation steel. *Computational Materials Science*, 78 (2013), 55–62.
11. C. Shiga, et al., Computer simulation of compressive residual stress in LTT welded joints of high strength steels. In *Proceedings of National Symposium on Welding Mechanics & Design*, Hampton, VA (2011).
12. J.E.M. Jubb, I.G. Phillips, and H. Becker, Interrelation of structural stability, stiffness, residual stress and natural frequency. *Journal of Sound & Vibration*, 39(1) (1975), 121–134.
13. W. Jiang, Y. Zhang, and W. Woo, Using heat sink technology to decrease residual stress in 316L stainless steel welding joint: Finite element simulation. *International Journal of Pressure Vessels and Piping*, 92 (2012), 56–62.
14. A. Ohta, et al., Fatigue strength improvement of lap joints of thin steel plate using low-transformation-temperature welding wire. *Welding Journal*, 82(4) (2003), 78.
15. A.A. Bhatti, et al., Fatigue strength improvement of welded structures using new low transformation temperature filler materials. *Procedia Engineering*, 66 (2013), 192–201.
16. N. Suzuki, A. Ohta, and Y. Maeda, Repair of fatigue cracks initiated around box welds using low transformation temperature welding material. *Welding International*, 18(2) (2004), 112–117.
17. A. Kromm, J. Dixneit, and T. Kannengiesser, Residual stress engineering by low transformation temperature alloys—state of the art and recent developments. *Welding in the World*, 58(5) (2014), 729–741.
18. C. Shiga, et al., Effect of Ms temperature on residual stress in welded joints of high-strength steels. *Welding in the World*, 54(3) (2010), R71–R79.
19. T. Hanji, et al., Fatigue strength of transverse attachment steel joints with single-sided arc weld using low transformation temperature welding consumable. *Welding in the World*, 64 (2020), 1293–1301.
20. M. Shirato, *Specifications for Highway Bridges*. Steel Construction Today & Tomorrow (2018).

21. Y. Mikami, et al., *Evaluation of Relationship between Transformation Expansion of Weld Metal and Welding Distortion through Numerical Simulation Considering phase Transformation Effect*. IIW Doc. X-1582-2005 (2005).
22. J. Lin, et al., Measurement of residual stress in arc welded lap joints by cosα X-ray diffraction method. *Journal of Materials Processing Technology*, 243 (2017), 387–394.
23. Y. Liu, F. Sommer, and E. Mittemeijer, Calibration of the differential dilatometric measurement signal upon heating and cooling; thermal expansion of pure iron. *Thermochimica Acta*, 413(1–2) (2004), 215–225.
24. M. Mochizuki, et al., Study on residual stress reduction by using phase transformation phenomena in welding material-studies on numerical simulation of temperature, microstructure, and thermal stress histories during welding process and their application to welded structures (Report 2). *Quarterly Journal of the Japan Welding Society*, 23(1) (2005), 112–121 (in Japanese).

4

Process Design by LTT Welding Materials for Fatigue Crack Prevention of Boxing Fillet-Welded Joints

4.1 Fatigue Strength of Boxing Fillet-Welded Joints

Figure 4.1.1 shows the relationship between the yield strength of a steel base plate and the fatigue strength of 2×10^6 cycles for base plates and welded joints of butt welds, non-load-carrying cruciform welds, and boxing fillet welds, which Watanabe and Matsumoto [1] obtained from the previous data on fatigue strengths of welded joints made from various high strength steels.

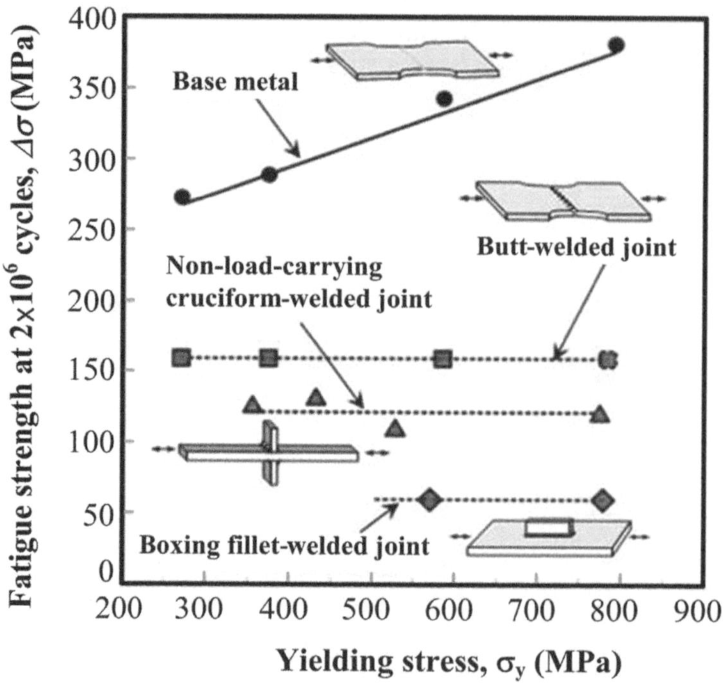

FIGURE 4.1.1
Relationship between yield strength of the base plate and fatigue strength for 2×10^6 cycles in some welded joints [1].

DOI: 10.1201/9781003601845-4

The increase in the yield strength of the base plate increases the fatigue strength; however, it does not increase the fatigue strength of the welded joints. In particular, the fatigue strength of boxing fillet-welded joint at the gusset front is the lowest among all. The main reason for the poor fatigue strength of high strength steel welded joints is thought to be the increase in tensile residual stress that accompanies the strengthening of steels.

There are many boxing fillet-welded joints in ship structures. Fatigue cracks often occur at the weld toe. To prevent and repair these fatigue cracks, surface dressing techniques to reduce stress concentration at weld toes [2–6] or peening techniques such as high-frequency impact treatment and ultrasonic impact treatment [7–10], and the low transformation temperature (LTT) welding techniques [11–14] have been developed to reduce residual stress.

Shiga et al. have verified the improvement in fatigue strength of fillet-welded joints using 10%Cr–10%Ni LTT welding material in comparison with conventional welding materials [15].

Figure 4.1.2 illustrates the size and shape of out-of-plane gusset (longitudinal stiffener) welded joints (boxing fillet-welded joints). The main plate thickness is 20 mm, and the gusset plate (stiffener) thickness is 16 mm. High strength steels (KA36) for ship structures of 490 MPa-class are used to fabricate the welded joints. Table 4.1.1 shows the chemical compositions of KA36 steel. Conventional and LTT welding materials, as shown in Table 4.1.2, were used for the robotic welding of out-of-plane gusset welded joints. The welding conditions are shown in Table 4.1.3.

Figure 4.1.3 shows the exteriors and macroscopic appearances of the weld beads. The fatigue testing results (cycles to failure, N_f) are listed in Table 4.1.4.

The relation between N_f and $\Delta\sigma$ of conventional boxing fillet-welded joints (A-Conv. joint) is shown in Figure 4.1.4.

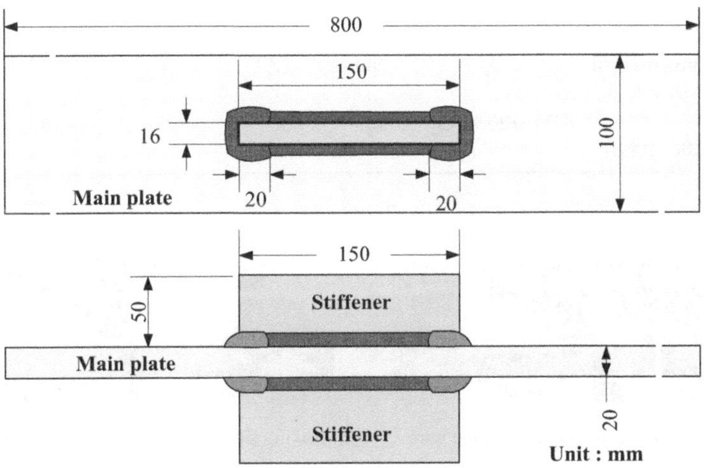

FIGURE 4.1.2
Fatigue test specimen of boxing fillet-welded joint.

TABLE 4.1.1

Chemical Compositions of High Strength Steels (KA36)

		Chemical Compositions (mass%)					Mechanical Properties			
Material	Thick. (mm)	C	S	Mn	P	S	YS (N/mm^2)	TS (N/mm^2)	EL (%)	$_VE_{0°C}$ (J)
KA36	20	0.14	0.23	1.08	0.04	0.006	446	543	21	235
KA36	16	0.14	0.21	1.1	0.017	0.004	454	511	23	256

TABLE 4.1.2

Chemical Compositions of Conventional and LTT Welding Materials

	Chemical Compositions (%)					
Filler Metals	C	Si	Mn	Ni	Cr	Fe
Conventional (MX-Z200)[a]	0.045	0.5	1.7	0.02	–	Balance
LTT (10Cr10Ni [Ohta]) [12]	0.025	0.32	0.7	10	10	Balance
LTT (10Cr10Ni [Shiga]) [17]	0.048	0.26	0.6	9.46	10.5	Balance
LTT (10Cr10Ni) [15]	0.048	0.26	0.6	9.46	10.5	Balance

[a] AWS A5.20 E70T-1C.

TABLE 4.1.3

Welding Conditions in Boxing Fillet Weld

Filler Material		Shielding Gas [25 L/min]	Current (A)	Voltage (V)	Speed (cm/min)	Heat Input (kJ/cm)	Remark
Fillet Weld	Conventional (MX-Z200)	CO_2	280	32	48	11	Robot
Boxing Fillet Weld	Conventional (MX-Z200)	CO_2	200	27.4	36	16	Robot
	LTT (10Cr10Ni)	80%Ar20%CO_2	225	30	25	16	Robot

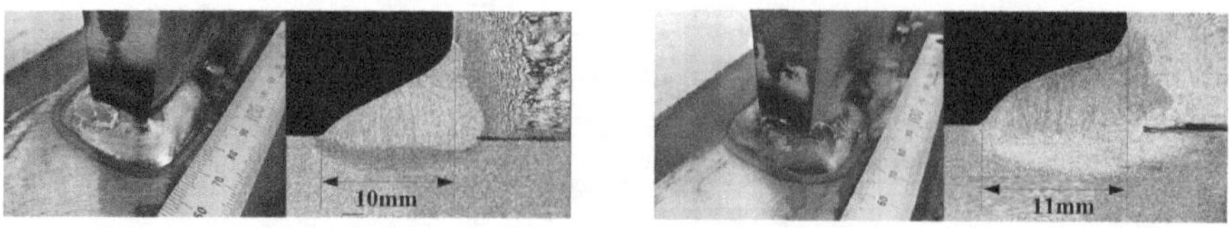

(a) Boxing fillet bead by conventional wire (b) Boxing fillet bead by LTT (10%Cr-10%Ni) wire

FIGURE 4.1.3
External appearance and cross section of welded joints.

TABLE 4.1.4

N_f Results in the Specimens of Types A, B, and AD

	Width	Thick	Cross-Section Area	Maximum Load	Stress Range	Number of Cycles to Failure	
Specimen No.	W (mm)	t (mm)	A (mm²)	P_{max} (kN)	$\Delta\sigma$ (N/mm²)	N_f (cycles)	Ratio of N_f of A-Conv.
A-Conv.-2	100.3	20.1	2,016	302	150	338,000	1.00
A-Conv.-3	100.2	20.2	2,024	303	150	353,000	(3.46×10^5)
A-Conv.-4	100.8	20.1	2,023	405	200	128,000	1.00
B-10Cr10Ni-2	99.8	20.3	2,023	304	150	298,000	0.86
B-10Cr10Ni-3	100.1	20.1	2,012	302	150	356,000	1.03
B-10Cr10Ni-4	98.6	20.2	1,992	398	200	138,000	1.08
AD-Conv.-1	98.8	20.4	2,016	302	150	691,010	2.00
AD-Conv.-2	98.1	20.4	2,001	300	150	834,275	2.41

500 kN hydraulic transmission fatigue testing machine, stress ratio R=0.0, and cycle speed 3–5 Hz.

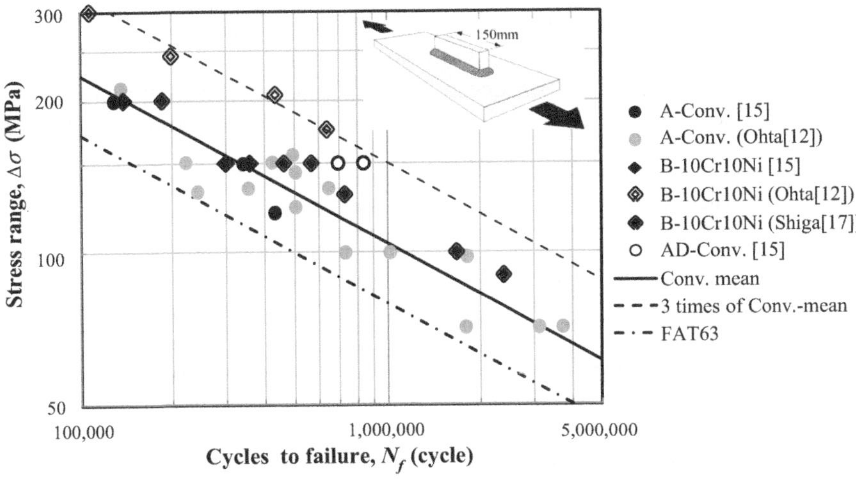

FIGURE 4.1.4
Fatigue test results of A-Conv., B-LTT, and AD welded specimens.

The fatigue resistance data is based on the number of cycles N to failure. The data is represented in S–N curves.

$$N = \frac{C}{\Delta\sigma^m}$$

Here, $\Delta\sigma$ is the stress range.

The slope of the fatigue strength curves for details assessed on the basis of normal stresses is $m = 3$ [16].

The regressed mean curve for $m = 3$ of A-Conv. joints is

$$\Delta\sigma = 10,400 N_f^{-1/3}$$

$$N_{f150\,\mathrm{MPa}} = 3.33\times10^5,\quad \Delta\sigma_{2e6} = 82.5\,\mathrm{MPa}$$

The International Institute of Welding (IIW) recommends a fatigue strength of 63 MPa for out-of-plane gusset joints at 2×10^6 cycles (FAT63[16]) in gusset plates of 150 mm length, as shown in Figure 4.1.5. FAT63's design (–2 STDV) curve is shown in Figure 4.1.4. It is shown that the fatigue strength of A-Conv. joints is almost equivalent to FAT63.

In all A-Conv. joints treated by toe grinding (AD-joint), fatigue cracks initiated at the root gap. After propagating through the boxing fillet weld and the main plate, they led to fracture. The fatigue strength of AD-joints was improved to 2–2.5 times compared with the fatigue life of conventional A-Conv. joints. Figure 4.1.6 shows an example of fracture surfaces.

All of the 10%Cr–10%Ni [15] and 10%Cr–10%Ni [17] LTT boxing fillet-welded joints (B-LTT joint) failed due to fatigue cracking at the weld toe on the base plate in the same manner as the conventional A-Conv. joints. This shows that the fatigue strength cannot be improved by simply using an LTT welding material without a change in the welding process.

In B-10Cr10Ni [12] joints processed with three-pass welding, the fatigue life of the joints improved to three times that of conventional joints.

No.	Structural Detail	Description	FAT Steel
521	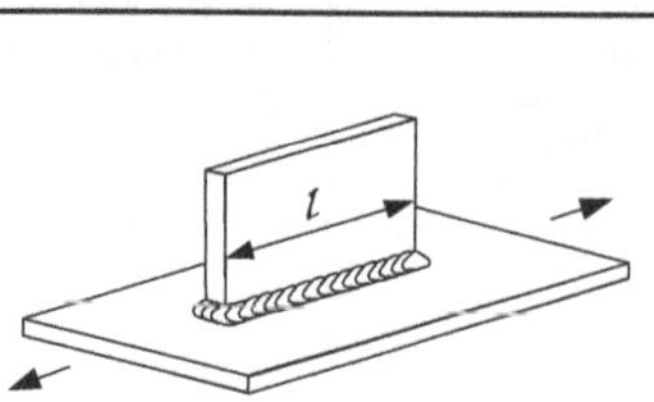	Longitudinal fillet welded gusset at length l l < 50mm l < 150mm l < 300mm l > 300mm	 80 71 63 50

FIGURE 4.1.5
Fatigue resistance values for structural details in steel assessed based on nominal stresses.

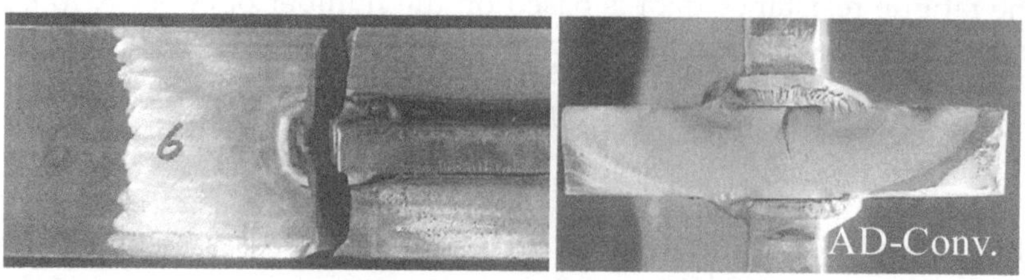

FIGURE 4.1.6
Fractures of AD specimen dressed at the conventional weld toe by grinder.

However, because fatigue cracks initiated at the weld toe on the base plate, it is difficult to achieve sufficient compressive stress at the toe of the base plate.

4.2 Concept of LTT Elongated Bead Welding Method

The conventional use of LTT weld has much less improved the fatigue strength than other impact treatment methods. This is because the reduction of residual stress in LTT weld is likely to be smaller than that achieving with the other methods.

Figure 4.2.1 shows the results of residual stress distributions measured by the neutron diffraction method [18]. The measurements were carried out at a depth of 2 mm from the surface along the measured line, as shown in Figure 4.2.1a.

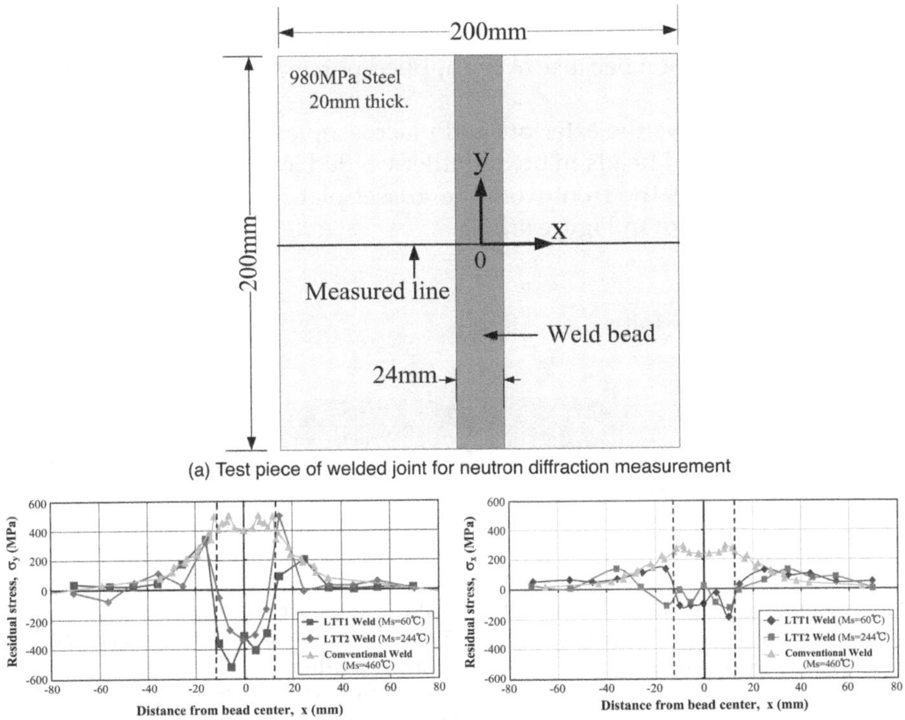

(a) Test piece of welded joint for neutron diffraction measurement

(b) In case of residual stress in y-direction (σ_y)

(c) In case of residual stress in x-direction (σ_x)

FIGURE 4.2.1
Comparison of residual stress distribution between conventional weld metal and low Ms weld metals (LTT1, LTT2).

Figure 4.2.1b and c shows the residual stress distributions along the measured line for the longitudinal stress component σ_y and the transverse stress component σ_x in butt welds using LTT welding materials, respectively.

The distribution of σ_x in butt welds corresponds to the stress distribution from the stiffener edge to the end of a fillet weld. It is noteworthy that, compared to the minimum value of −200 MPa in σ_x, σ_y generates an extremely high compressive residual stress of −600 MPa, which is comparable to or even greater than that achieved by high-frequency impact treatment or ultrasonic impact treatment.

The large compressive residual stress of σ_y in LTT weld metal can be utilized to improve fatigue strength. To enhance the compressive residual stress and increase the resistance to the initiation and propagation of cracks (hot spots), it is important to choose the welding direction transverse to the crack direction, that is, the external loading direction, and to determine the elongated bead length in proportion to the required compressive residual stress.

The schematic drawing representing the concept of the LTT elongated bead welding method is shown in Figure 4.2.2, where the applicable portion (hot spot) and welding procedure of the elongated bead weld are summarized [19]. When applying the elongated bead weld to hull structures, it is preferable to adjust the elongated bead length in proportion to the restraint force of the applied portion because every applied portion has a different restraint force.

Figure 4.2.3 shows the exteriors and macroscopic appearances of the conv. and elongated weld beads in boxing fillet-welded joints. The elongated bead is overlaid towards the front from the gusset plate end on the conv. boxing fillet weld, as shown in Figure 4.2.3b.

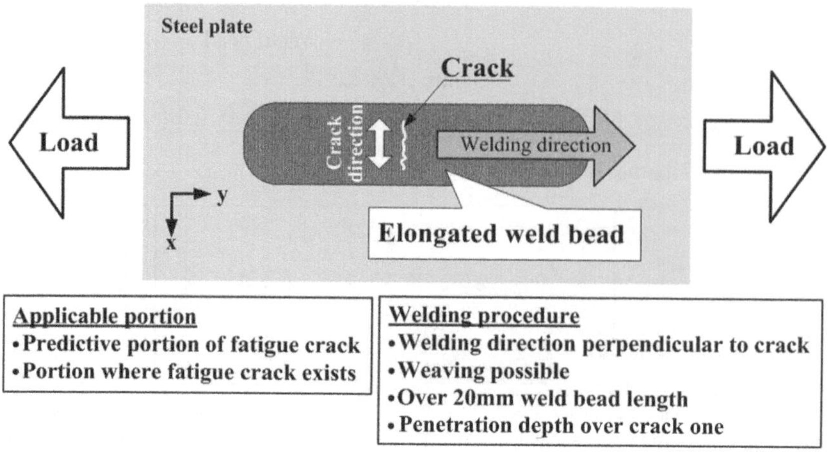

FIGURE 4.2.2
Schematic drawing representing the concept of the elongated weld bead method, welding procedure, and applicable portion.

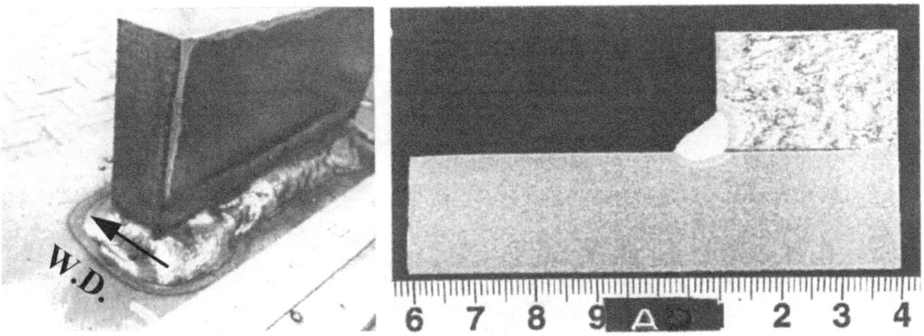

(a) Conventional weld in boxing fillet-welded joints

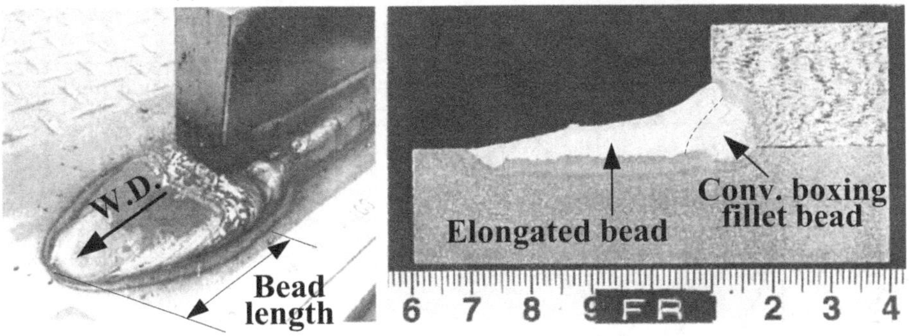

(b) Elongated weld in boxing fillet-welded joints

FIGURE 4.2.3
External appearance of boxing fillet-welded joints.

The effect of the length of the elongated bead welded using LTT wire on the longitudinal residual stress (σ_y) was computed by finite element (FE) analyses. In the case of the LTT elongated bead with Ms=200°C, the residual stress distributions on the surface along the line *L* are shown in Figure 4.2.4 [19].

In all cases of elongated bead lengths of 20, 40, and 70 mm, significant compressive residual stresses are generated on the bead surface. The compressive residual stress increases with longer bead lengths but saturates at approximately −600 MPa when the bead length reaches 40 mm or more. Similarly, the ends of the elongated beads also exhibit saturation with high compressive residual stress.

Based on the above, in gusset welded joints, it can be expected that a compressive residual stress of significant magnitude will be achieved if the elongated bead length is set to 40 mm or more.

From Figure 4.2.3, it is confirmed that the angle of the elongated bead toe with the base plate is much larger than that of usual welds, which substantially reduces the local stress concentration at the weld toe. The change in maximum principal stress along line *L* from the gusset end to the base plate

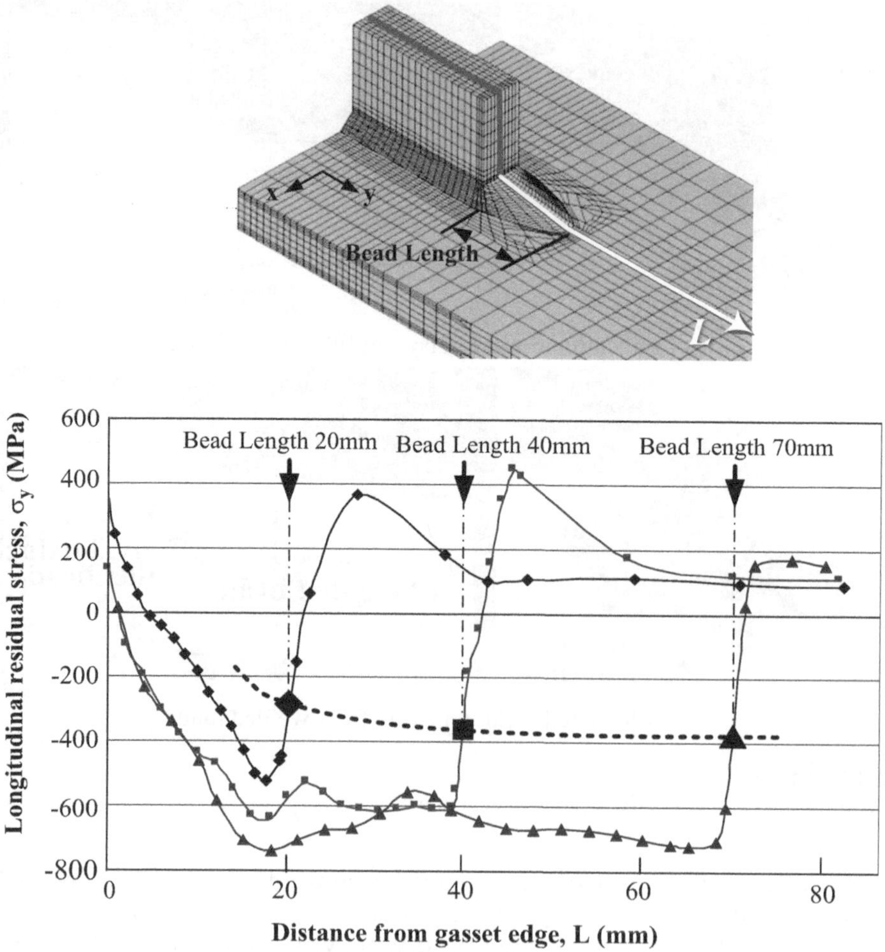

FIGURE 4.2.4
Effect of elongated bead length on residual stress distribution.

side, as shown in Figure 4.2.5, was calculated in the model of the elongated bead weld under an applied stress of 150 MPa [19].

For elongated bead lengths of 20, 40, and 70 mm, the peak stresses are 304, 230, and 198 MPa, respectively, and these are observed at the plate-side toe position. The stress concentration increases slightly at the gusset-side toe with an increase in elongated bead length, but the stress concentration is reduced at the plate-side toe position.

The elongated bead length has a beneficial effect on reducing stress concentration under external fatigue stress as well as on the formation of compressive residual stress at the plate-side toe position.

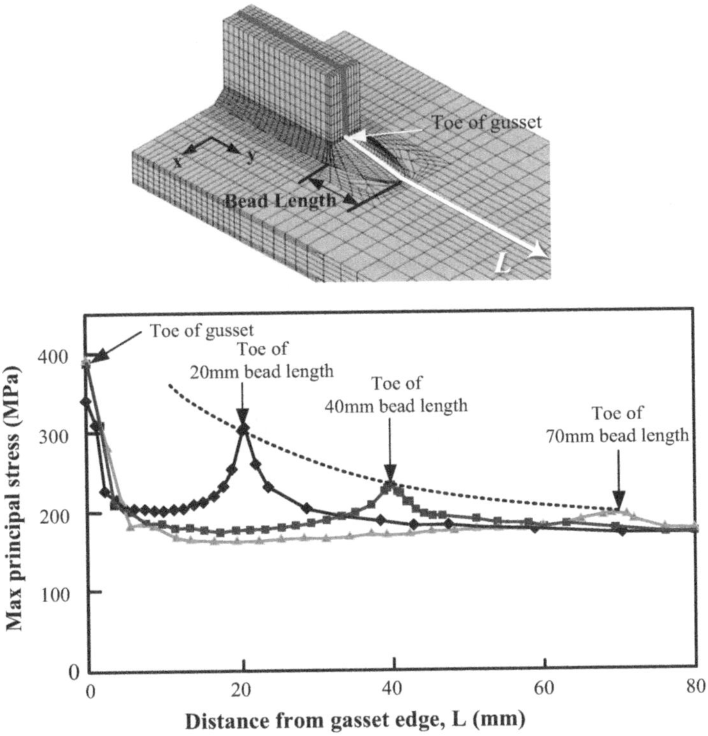

FIGURE 4.2.5
Effect of elongated bead length on stress concentration at the toe of the plate under an applied stress of 150 MPa.

4.3 Fatigue Characteristics of Boxing Fillet-Welded Joint with LTT Elongated Bead

4.3.1 Elongated Bead with 10%Cr-10%Ni LTT Welding Material

The fatigue strength improvement effect and the applicability of LTT elongated bead welding for ship's welded joints are examined [15,20]. The main plate thickness is 20 mm, and the gusset plate (stiffener) thickness is 16 mm. High strength steel for ship structures of 490 MPa-class is used to fabricate the welded joint.

Figure 4.3.1 shows a detailed drawing of the welding procedure for the boxing fillet weld specimen. Fillet welds are formed with conventional welding wire (Kobelco: MX-Z200) (Figure 4.3.1a), and boxing fillet welds are also formed with conventional welding wire (Figure 4.3.1b: Conv. boxing fillet-welded joint). Conv. boxing fillet and elongated bead welds are performed using robotic welding.

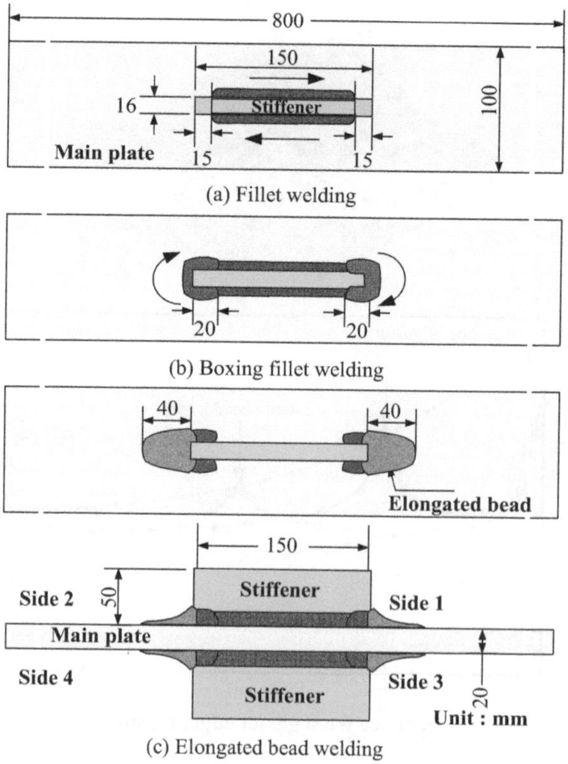

FIGURE 4.3.1
Size and shape of the out-of-gusset welded joint specimen with an elongated bead.

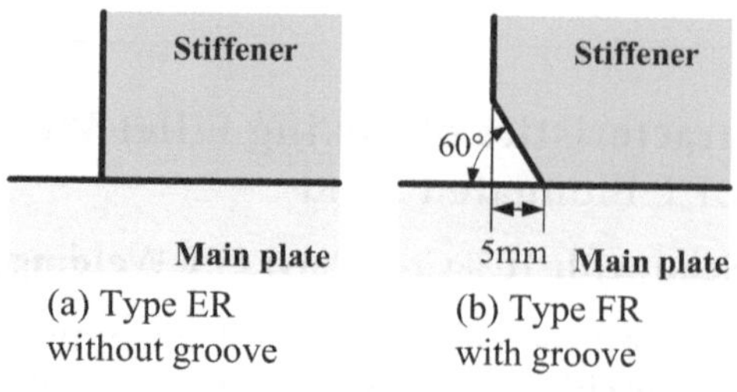

FIGURE 4.3.2
Fatigue test specimen featuring a boxing fillet-welded joint with or without a groove.

In specimens of type ER, shown in Figure 4.3.2a, an elongated bead weld with a 10%Cr–10%Ni LTT welding wire (Figure 4.3.1c) was performed using a weaving technique. The length of this elongated bead was 40 mm. The elongated beads are fabricated sequentially from Side 1 to Side 4.

In specimens of type FR, a groove was processed at an angle of 60° and a depth of 5 mm at the bottom of the gusset plate (Figure 4.3.2b). Fillet and boxing fillet welds with conventional welding material, as well as an elongated bead weld with an LTT welding material (10%Cr–10%Ni), the same as in "ER", were performed.

All tests were conducted in air at room temperature, maintaining a constant amplitude stress ratio, $R=0$, with load control and frequencies ranging from 3 to 5 Hz. The nominal stress range $\Delta\sigma$ was between 150 and 200 MPa.

The fatigue testing results (cycles to failure, N_f) are listed in Table 4.3.1. The relation between N_f and $\Delta\sigma$ for FR-10Cr10Ni and ER-10Cr10Ni joints is shown in Figure 4.3.3.

TABLE 4.3.1

Fatigue Testing Results of ER and FR Specimens with LTT Elongated Bead

No.	Specimen Type	$\Delta\sigma$ (MPa)	N_f (Cycles)
1	FR-10Cr10Ni	150	1,410,000
2	FR-10Cr10Ni	150	1,450,000
3	FR-10Cr10Ni	150	2,150,000
4	FR-10Cr10Ni	150	1,880,000
5	FR-10Cr10Ni	200	1,340,000
6	FR-10Cr10Ni	200	1,290,000
7	ER-10Cr10Ni	150	3,233,000
8	ER-10Cr10Ni	200	1,067,000
9	ER-10Cr10Ni	150	2,528,000
10	ER-10Cr10Ni	200	761,000
11	ER-10Cr10Ni	150	1,196,000

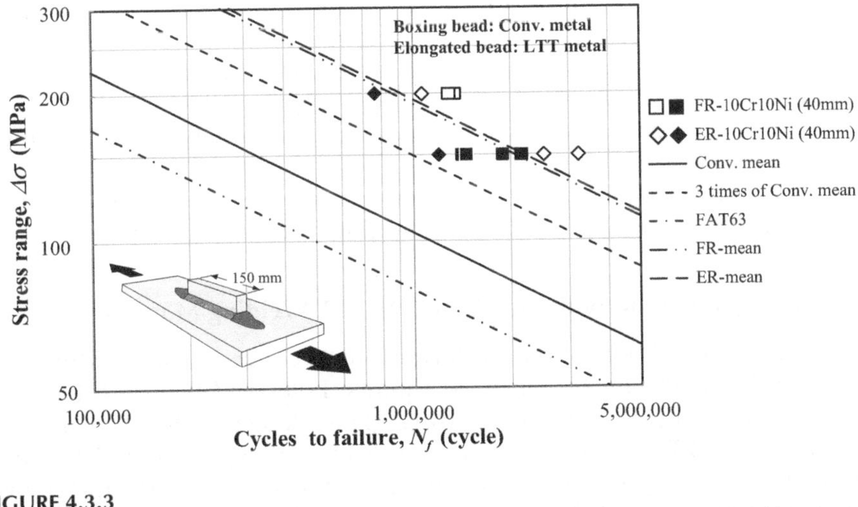

FIGURE 4.3.3
SN diagram of fatigue testing results of ER and FR specimens with LTT elongated bead.

For FR and ER processed joints, the regressed mean curves and the characteristic parameters are

$$\Delta\sigma = 19{,}055 \times N_f^{-1/3}$$

$$N_{f150\,\mathrm{MPa}} = 2.05 \times 10^6, \quad \Delta\sigma_{2e6} = 151\ \mathrm{MPa}$$

for FR-10Cr10Ni joints, and

$$\Delta\sigma = 19{,}498 \times N_f^{-1/3}$$

$$N_{f150\,\mathrm{MPa}} = 2.20 \times 10^6, \quad \Delta\sigma_{2e6} = 155\ \mathrm{MPa}$$

for ER-10Cr10Ni joints.

These results show that the characteristic parameters of the two joints are almost identical. The averaged ratios $\Delta\sigma_{2e6}^{(\mathrm{LTT})}/\Delta\sigma_{2e6}^{(\mathrm{Conv})}$ and $N_{f150\,Mpa}^{(\mathrm{LTT})}/N_{f150\,Mpa}^{(\mathrm{Conv})}$ are 185% and 638%, representing about an 85% increase in mean fatigue strength and a 6.4 times longer failure life. It has been shown that a significant improvement in fatigue strength is achieved by making an elongated bead with an LTT wire [20].

It is also shown that whether or not a bevel is made on the gusset plate does not affect the fatigue strength in these cases.

In all ER-10Cr10Ni and FR-10Cr10Ni joints marked with solid symbols (■, ◆) in Figure 4.3.3, fatigue cracks initiated at the root gap. After propagating through the boxing weld and the main plate, this led to a fracture. Figure 4.3.4a and c shows examples of fracture appearance and fracture surface.

In all ER-10Cr10Ni and FR-10Cr10Ni joints shown by open symbols (□, ◇), fatigue cracks initiated at the boundary between fillet bead and boxing bead. After the cracks propagated through both the gusset plate and main plate, this led to a fracture. Figure 4.3.4b and d shows examples of fracture appearance and fracture surface. In the boundary failure mode, the length of the crack propagation path becomes significantly longer than that in the weld root cracking mode. Therefore, the fatigue strength in the boundary failure mode was improved to 6–6.5 times or more compared with the fatigue life of conventional A-Conv. joints.

However, it is thought that the presence of the two failure modes caused the fatigue life to scatter significantly, ranging from three to ten times.

4.3.2 Elongated Bead with 16%Cr–8%Ni LTT Welding Material

Fillet and boxing fillet welds are formed with conventional welding wire (Kobelco: MX-Z200) for ER processed joints. Fillet and boxing fillet welds are performed using robotic welding. The elongated bead weld with 16%Cr–8%Ni LTT welding wire (Kobelco: DW-168FP) is formed by semi-automatic arc welding rather than robotic welding [21]. The chemical compositions of 16%Cr–8%Ni LTT wire are shown in Table 4.3.2.

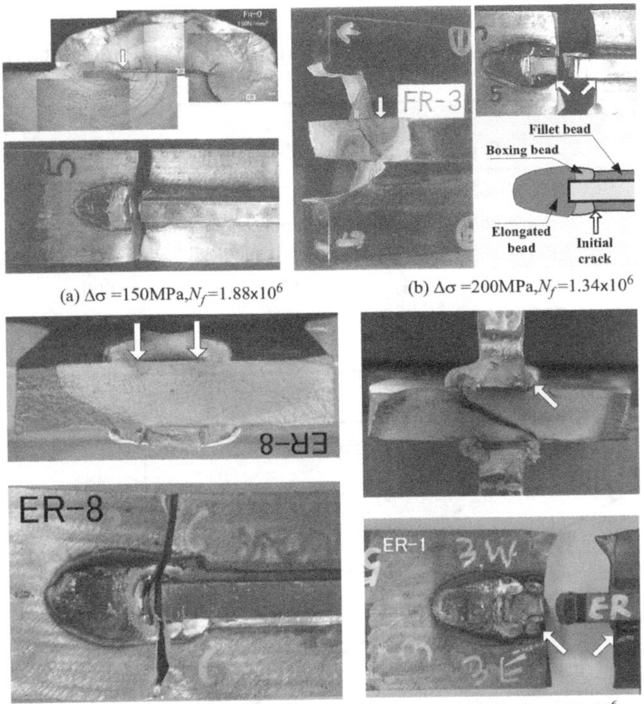

(a) $\Delta\sigma$ =150MPa,N_f=1.88x10^6 (b) $\Delta\sigma$ =200MPa,N_f=1.34x10^6 (c) $\Delta\sigma$ =150MPa,N_f=1.20x10^6 (d) $\Delta\sigma$ =150MPa,N_f=3.23x10^6

FIGURE 4.3.4
Fractures of boxing fillet-welded joints with 10%Cr–10%Ni LTT elongated bead.

TABLE 4.3.2
Chemical Compositions of 16%Cr–8%Ni LTT Deposit Metal

	Chemical Compositions (Mass%)							Mechanical Properties			
Wire	**C**	**Si**	**Mn**	**P**	**S**	**Ni**	**Cr**	**0.2%PS (N/mm^2)**	**TS (N/mm^2)**	**EL (%)**	**vE0°C (J)**
16Cr8Ni LTT (DW-168FP)	0.02	0.34	0.44	0.02	0.01	8.2	16.1	321	845	19	31

The fatigue testing results are listed in Table 4.3.3. The relation between N_f and $\Delta\sigma$ for ER-16Cr8Ni joints is plotted with solid marks in Figure 4.3.5.

All test results indicate that fatigue cracks initiated from the root. The fatigue life extension of ER-16Cr8Ni LTT specimens is limited to less than three times compared with the conventional boxing fillet welds.

This is because the penetration depth of ER-16Cr8Ni LTT specimens is almost zero compared with the 3mm penetration depth in ER-10Cr10Ni LTT specimens, as shown in Figure 4.3.6.

Then, to ensure reliable penetration, an additional test specimen of the FR type with an edge groove was added. To fabricate the FR type test specimen,

TABLE 4.3.3
Fatigue Test Results of Boxing Fillet-Welded Joint with Elongated Bead by Using 16%Cr–8%Ni LTT Wire

Specimen No.	Type	$\Delta\sigma$ (MPa)	N_f (Cycles)	Ratio for Conv. Mean
1	ER-16Cr8Ni	150	896,000	2.6
2	ER-16Cr8Ni	150	716,000	2.1
3	ER-16Cr8Ni	150	691,000	2.0
4	GR-16Cr8Ni (a=5)	150	1,720,000	5.2
5	GR-16Cr8Ni (a=5)	150	1,770,000	5.3
6	GR-16Cr8Ni (a=5)	150	>3,620,000	>10.5

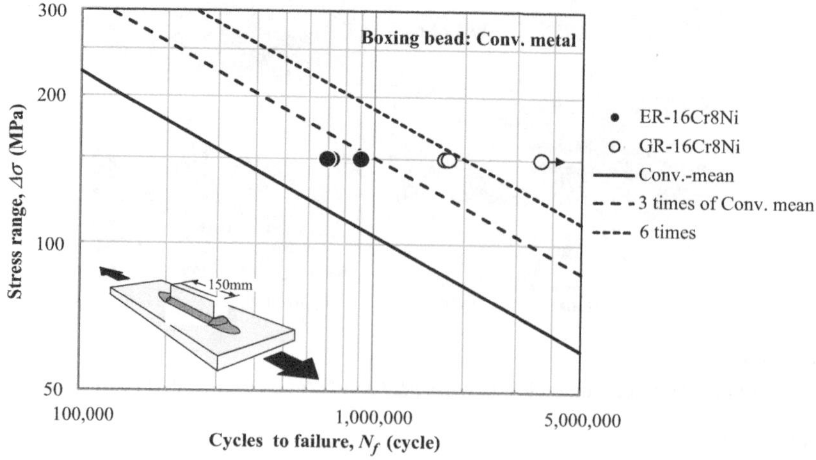

FIGURE 4.3.5
SN diagram of fatigue test results of ER and GR specimens with an elongated bead using 16%Cr–8%Ni LTT wire.

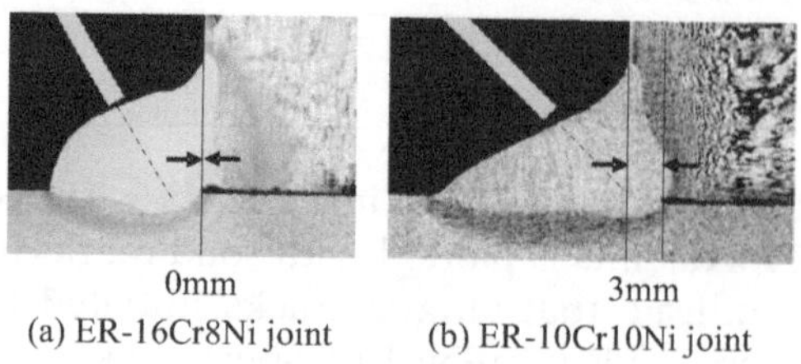

FIGURE 4.3.6
Effect of welding conditions on penetration depth at stiffener end.

as shown in Figure 4.3.7, the air arc gouging process was employed, and a groove depth of 5 mm was prepared before boxing fillet and elongated bead welding.

FIGURE 4.3.7
Procedure for obtaining penetration at the stiffener end.

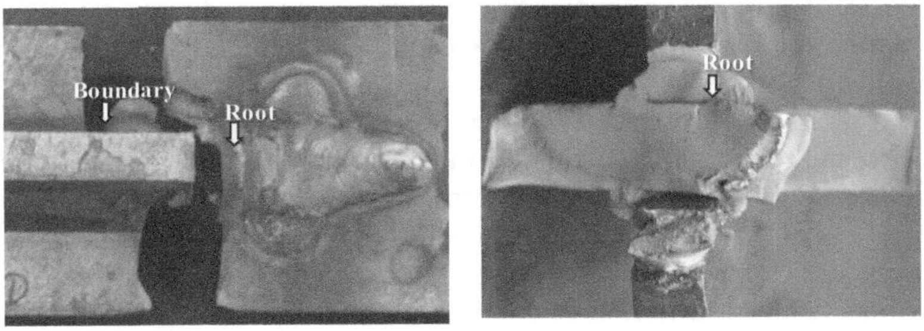

FIGURE 4.3.8
Fractures of GR-16Cr8Ni LTT specimens.

Subsequently, the boxing fillet weld was re-formed using semi-automatic arc welding with conventional wire. After this, the elongated bead welding was applied using a 16%Cr–8%Ni LTT wire. This test specimen was named the GR type.

The fatigue testing results are appended in Table 4.3.3. The relation between N_f and $\Delta\sigma$ of GR-16Cr8Ni joints is plotted with open marks in Figure 4.3.5.

The fatigue life extension for GR-16Cr8Ni LTT elongated bead test specimens was 5 to over 10 times higher than the conventional one, showing nearly the same improvement as ER-10Cr10Ni and FR-10Cr10Ni LTT elongated beads.

This suggests that penetration into the gusset edge significantly contributes to delaying fracture initiation from the root.

The fractured specimen ER-16Cr8Ni-4 and the fracture surface are shown in Figure 4.3.8. Although the crack initiated from the root, cracks propagated from the boundary between the straight fillet bead and the boxing fillet bead, similar to Figure 4.3.4b and d. This suggests that the fracture is in a transition region between root fracture and boundary fracture.

4.4 Prevention of Root Crack in Boxing Fillet-Welded Joints

The fatigue life of LTT elongated bead specimens can be extended by 2 to more than 6 times compared to conventional boxing fillet-welded joints. However, the large variation range is an issue for practical applications.

If conditions to suppress crack initiation and propagation from the root can be effectively controlled, it would be possible to reduce the variation range and achieve a life extension of six times or more.

Figure 4.4.1 is a diagram defining the penetration parameters *a* and *b* at the root of the boxing fillet-welded joint. If the root is considered to correspond to the crack tip, the crack propagation behavior under cyclic loading (stress range $\Delta\sigma = 150\,\text{MPa}$) can be numerically analyzed using the FEM program "JWRIAN" developed by Osaka University [22].

Firstly, to examine the effect of cyclic loading, the residual stress field was ignored. Additionally, the numerical analysis was conducted on the welded specimens with a main plate thickness of 20 mm and a stiffener plate thickness of 16 mm.

Figure 4.4.2 shows the crack propagation path. At $a = 5\,\text{mm}$ and $b = 2\,\text{mm}$, the crack propagates into the stiffener, indicating that the fracture path becomes longer. Meanwhile, Figure 4.4.3 illustrates the crack growth rate. As *a* and *b* increase, the number of cycles (load cycles) required to reach a certain length increases, indicating that the propagation slows down.

To simplify the evaluation of the stress intensity factor at the root region, the root was treated as the crack tip, and numerical analysis was conducted. The relationship between the stress intensity factor *K* and the penetration parameters *a*, *b* can be summarized in Figure 4.4.4. The crack propagation length and speed at different penetration parameters *a* and *b* can be described by Cases (a), (b), (c), (d), and (e) in Figure 4.4.3. It can be observed that the

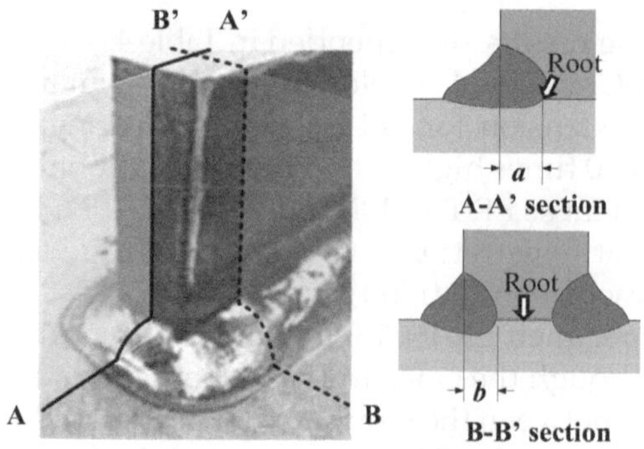

FIGURE 4.4.1
Definition of penetration parameters *a* and *b* at the boxing fillet weld.

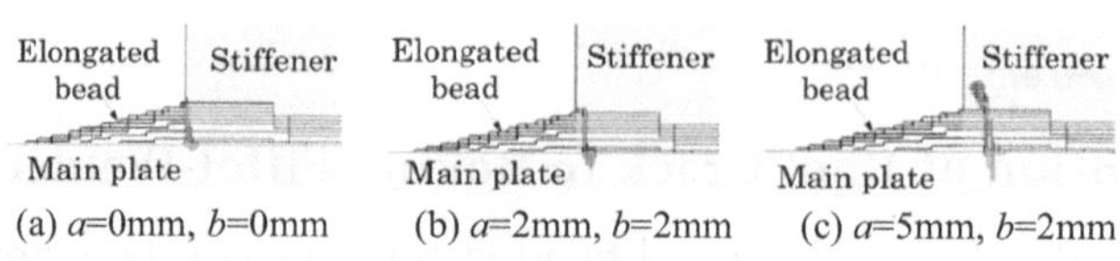

FIGURE 4.4.2
Crack propagation course under repeated loading.

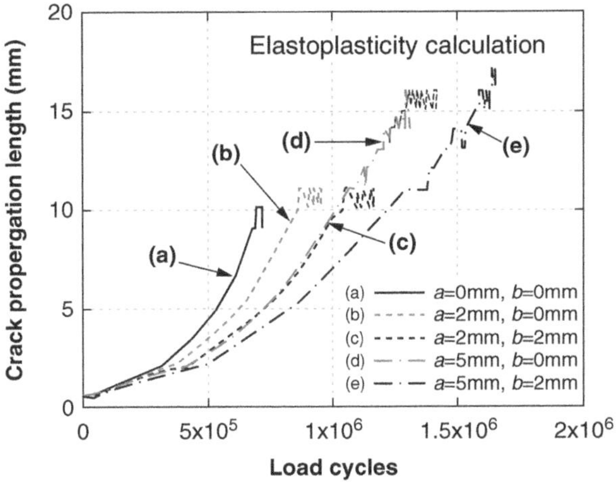

FIGURE 4.4.3
Crack propagation speed under repeated loading.

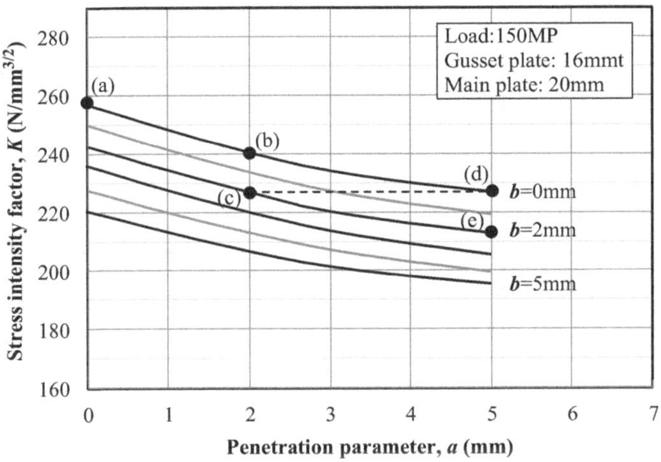

FIGURE 4.4.4
Influence of penetration parameters a and b on the stress intensity factor K at an average stress of 150 MPa.

propagation speed decreases as the stress intensity factor K reduces. This indicates that the stress intensity factor exhibits a strong correlation with fatigue life.

Based on the above results, it can be concluded that increasing the root penetration depths a and b is effective in extending fatigue life.

In reality, a tensile residual stress field exists in the root area, which may promote fatigue crack propagation. Figure 4.4.5 shows the numerically analyzed results of the residual stress field in the root area for cases $a=0$ mm and $a=5$ mm.

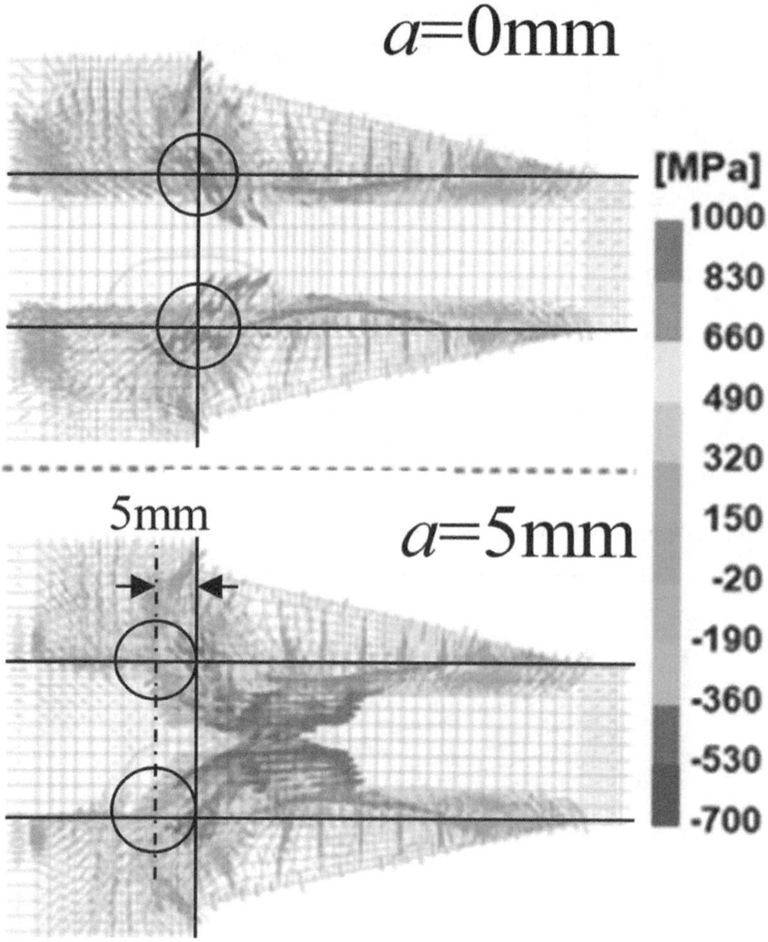

FIGURE 4.4.5
Residual stress distribution at the bead root of the stiffener end.

As parameter *a* increases, the tensile residual stress tends to decrease, further contributing to the extension of fatigue life.

In the previous section, the fatigue test results of LTT specimens ER-10Cr10Ni, FR-10Cr10Ni, ER-16Cr8Ni, and GR-16Cr8Ni were verified. An example of the measured penetration parameter *a* for various specimen types is shown in Figure 4.4.6. Additionally, parameter *b* ranges from 0.5 to 1.5mm for all specimen types, without significant differences. These results are plotted in a relationship diagram between the stress intensity factor *K* and parameters *a* and *b*, as shown in Figure 4.4.7.

The test results for the ER-16Cr8Ni specimen show a stress intensity factor *K* of approximately 250 $N/mm^{3/2}$, while the ER-10Cr10Ni, FR-10Cr10Ni, and

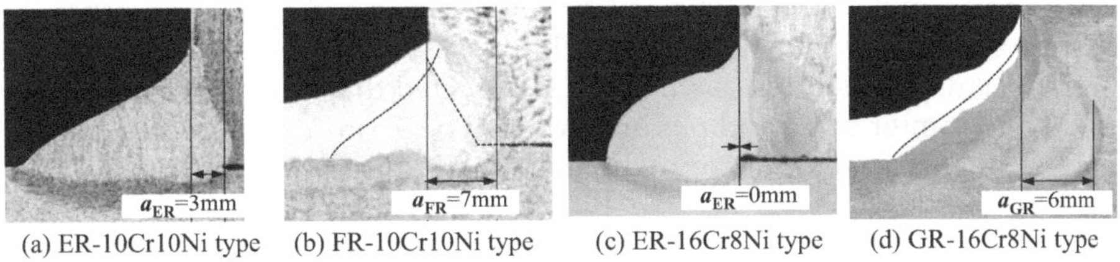

FIGURE 4.4.6
Measured penetration parameter ***a*** in each specimen type.

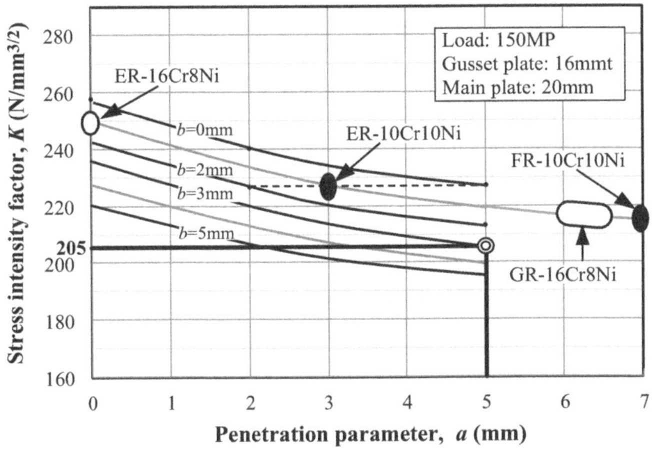

FIGURE 4.4.7
Diagram of the stress intensity factor *K* vs penetration parameters plotting measured *a* and *b* for all specimen types.

GR-16Cr80Ni specimens can be categorized with stress intensity factors *K* of around 220 N/mm$^{3/2}$.

As shown in Figures 4.3.3 and 4.3.5, the fatigue life extension of ER-16Cr8Ni is 2–3 times that of the conventional life, and all these specimens failed at the root. In contrast, the fatigue lives of ER-10Cr10Ni, FR-10Cr10Ni, and GR-16Cr80Ni are all 3–10 times or more longer than the conventional life. Moreover, in the high-life region exceeding about six times the conventional life, the failure mode shifts from root failure to fracture at the weld toe/bead boundary.

Therefore, it can be inferred that the reduction in the stress intensity factor *K* delays the initiation of root cracks. Thus, the stress intensity factor *K* can be considered a useful evaluation parameter for suppressing crack propagation from the root.

To further enhance the inhibitory effect against root failure, for a gusset plate thickness of 16 mm and a main plate thickness of 20 mm, the root shape with a=5 mm and b=3 mm (◎ mark in Figure 4.4.7), which reduces the stress intensity factor at the root to approximately K=205 N/mm$^{3/2}$ under a load of 150 MPa, shall be adopted as the basic design value.

These parameters, a and b, must be reliably ensured through mechanical means or arc gouging. After actual welding, since the penetration tends to be deeper than the basic design values, it is expected that, in practice, parameter a will be 6 mm or more and parameter b will be 3.5 mm or more, resulting in a stress intensity factor of approximately $K=200$ N/mm$^{3/2}$ or lower.

4.5 Repair and Reinforcement Procedures in All Welding Positions

The effectiveness of measures to prevent root crack initiation in the previous section is verified, and the examination of whether the elongated bead welding method can be applied as a practical fatigue life improvement technology in ship repair sites is conducted [23,24].

4.5.1 Procedure for Repair and Reinforcement Welding Specimens

A conventional boxing fillet weld specimen is prepared for a fatigue test as shown in Figure 4.1.2. The gusset plate thickness is 16 mm, and the main plate thickness is 20 mm.

The boxing fillet weld region of this test specimen is shown in Figure 4.5.1a.

When a fatigue crack occurs at the weld toe on the main plate side in a typical boxing fillet weld, the fatigue crack is completely removed using the air arc gouging method, as shown in Figure 4.5.1b.

At the same time, the boxing fillet weld around the stiffener edge is removed using the air arc gouging method while ensuring penetration depths a and b. In this process, the gouging length L of penetration b shall be 20 mm or greater.

The test specimen with this crack repair is referred to as the "Repaired Specimen: RP."

(a) Conventional boxing fillet-welded joint

(b) In joint with fatigue crack, gouging both fatigue crack part and boxing fillet bead.

(c) In joint without fatigue crack, gouging boxing fillet bead.

FIGURE 4.5.1
Arc gouging procedure for reinforced and repair fatigue test specimens.

Subsequently, if no fatigue cracks occur, the existing boxing fillet weld is removed using the air arc gouging method to secure penetrations *a* and *b*. The boxing fillet weld after gouging is shown in Figure 4.5.1c. This process is essentially a reinforcement method to strengthen the boxing fillet-weld joints in existing or new structures. Therefore, this test specimen is referred to as the "Reinforced specimen: RF". However, in the case of new fabrication, parameters *a* and *b* may be secured in advance by machining or other means.

From the perspective of such reinforcement and repair welding, unlike new construction, structures such as ships cannot be rotated for repair welding in a flat position, requiring welders to manually perform reinforcement and repair welding in all welding positions.

Therefore, when applying the LTT elongated bead welding method to practical applications, the following requirements must be satisfied.

(Basic rule)

1. It must be applicable in all welding positions (flat, horizontal, vertical-upward, and overhead).
2. It must be executable with semi-automatic arc welding by welders who are generally proficient in welding operations and must not require special welding techniques or tools.
3. It must be operable in confined spaces within existing hull structures. Specifically, boxing fillet welding should be designed to be performed by welding each side separately, considering cases where the welder cannot position themselves directly in front of the workpiece.

Based on these fundamental requirements, the basic procedures for the reinforcement and repair test specimens using the "LTT Elongated Bead Welding Method" after gouging are established as follows in steps (1)–(4).

When repair of fatigue cracks is required, the process begins at step (1); for reinforcement only, it starts at step (2).

1. Repair welding on the fatigue crack removal area shall be performed using conventional welding wire (Figure 4.5.2a). Subsequently, non-destructive testing shall be conducted to confirm the presence or absence of defects.
2. The removed fillet weld shall be rewelded using conventional welding wire (Figure 4.5.2b).
3. Elongated bead welding shall be carried out using 16%Cr–8%Ni LTT flux-cored wire (FCW). The length of the elongated bead shall be 45–50 mm or more (Figure 4.5.2c).
4. In the case of vertical-up or overhead welding from the toe of the elongated bead, the penetration at the weld start point (weld toe)

tends to be shallow. Therefore, removing 5mm from the weld toe portion of the elongated bead by grinder is recommended as shown in Figure 4.5.2d. To avoid misoperations in the shipyard, the toe portion of elongated beads welded in all welding positions is removed by about 5mm using a grinder.

16%Cr–8%Ni LTT FCW (Kobelco: DW-168FP) with 100% CO_2 shielding gas is employed for elongated bead welding in all positions. Conventional wire (Kobelco: DW-100) is used for re-boxing fillet welding.

The chemical compositions and mechanical properties of the conventional wire are presented in Table 4.5.1.

The detailed procedure for elongated bead and boxing fillet welding in each welding position is described below. Note that these procedures are common to both RP specimens and RF specimens.

4.5.2 Welding Process for Elongated Bead in All Welding Positions

The appearance of LTT elongated beads welded by semi-automatic CO_2 arc welding in various welding positions is shown in Figure 4.5.3.

The flat welding position is described by (*F*) in Figure 4.5.3a, and continuous weaving welding is used.

In other welding positions, intermittent welding with arc ON/OFF and weaving is employed to prevent molten metal from dropping. Additionally, for the horizontal position (*H*) in Figure 4.5.3b and the vertical-upward

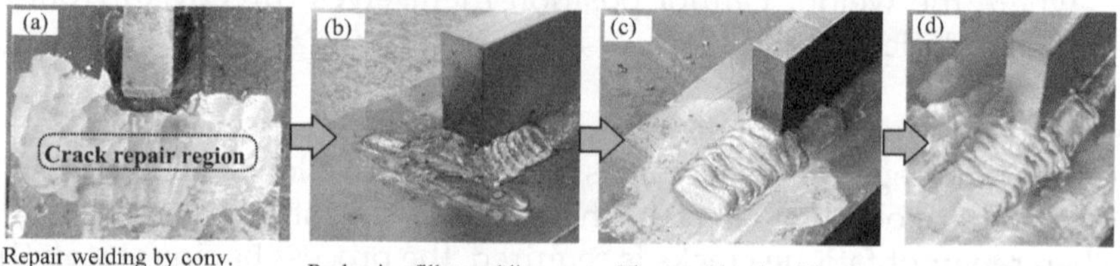

FIGURE 4.5.2
Welding procedure for the elongated bead method for reinforced and repair fatigue test specimens.

TABLE 4.5.1
Chemical Compositions and Mechanical Properties of Welding Materials

	Chemical Compositions (Mass%)							Mechanical Properties			
Wire	C	Si	Mn	P	S	Ni	Cr	0.2%PS (N/mm²)	TS (N/mm²)	EL (%)	vE0°C (J)
Conv.[a] (DW-100)	0.05	0.45	1.35	0.013	0.009	–	–	510	570	26	110

[a] AWS A5.20 E71T-1C.

progression positions (VU/U and VU/L) in Figure 4.5.3d and e, intermittent welding is also employed by layering beads from the bottom to the top. For the overhead position (OH) in Figure 4.5.3c, intermittent welding is carried out from the toe of the elongated bead toward the stiffener edge.

Standard welding conditions for each welding position are provided in Table 4.5.2.

The LTT elongated bead exhibits a smooth profile in the flat welding position. However, in other welding positions, the surface becomes uneven due to the intermittent welding process.

4.5.3 Process for Boxing Fillet Welding in All Welding Positions

Regarding to the groove after gouging treatment and its re-welding in various positions, the basic building-up procedures and welding direction are proposed as shown in Figure 4.5.4 based on preliminary experiments of semi-automatic arc welding.

As described above, the intermittent welding process is applied to elongated bead welding in the horizontal, vertical, and overhead positions, while a continuous weaving welding process is used in the flat welding position. The recommended welding directions are also shown by arrows in Figure 4.5.3.

Re-boxing fillet welding is performed on each side separately using conventional welding wire with multiple layers in all welding positions.

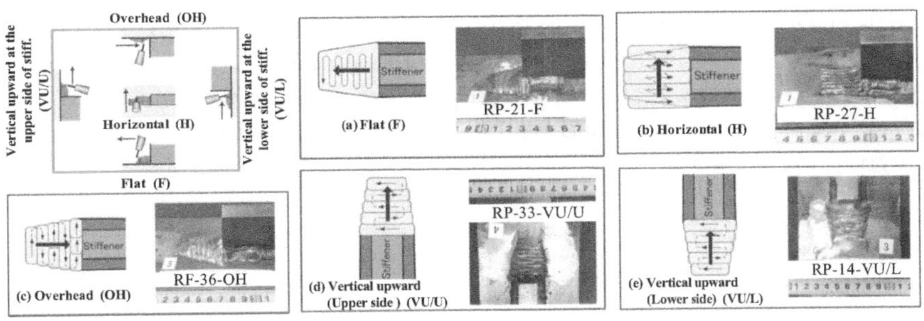

FIGURE 4.5.3
Welding operation of LTT elongated bead in each welding position.

TABLE 4.5.2
Welding Conditions for LTT Elongated Bead in Each Welding Position

Welding Position	Current (A)	Voltage (V)	Arc Time Ratio (%)	Welding Time (s)
Flat	180	22.4	100	33
Horizontal	180	26.5	80	63
Overhead	186	26.5	70	47
Vertical upward	183	26.2	80	57

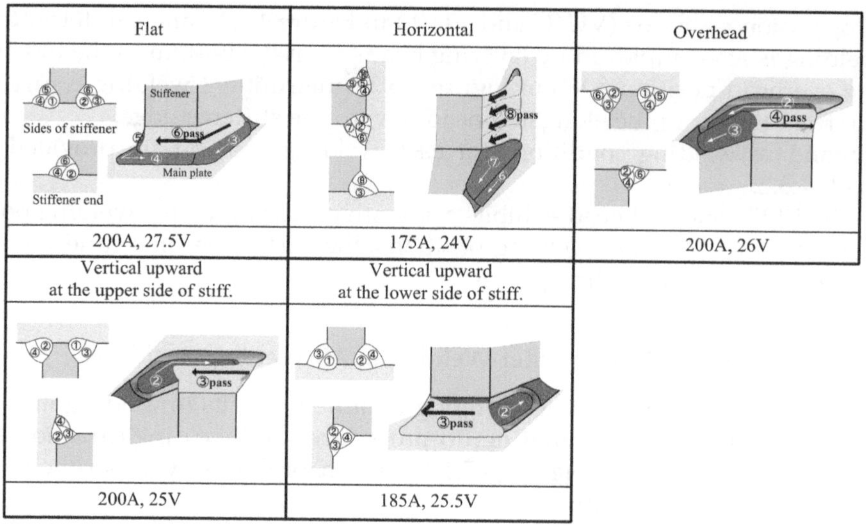

FIGURE 4.5.4
Welding procedures for the boxing fillet welds in various welding positions.

This meets the basic conditions for welding operations at the ship repair site.

In multi-layer welding, slag removal between each pass is essential for defect-free joints. Particularly in the overlay welding of gouged grooves, weld defects such as slag inclusions and lack of fusion are prone to occur, requiring careful attention. Detailed procedures for multi-layer welding to ensure defect-free joints are specified in the "Welding Guidelines". One example is shown in Figure 4.5.4.

4.6 Numerical Analysis of Residual Stress in LTT Elongated Bead in All Welding Positions

4.6.1 Residual Stress Analysis Modelling

The residual stress distribution in boxing fillet-welded joint is influenced by various factors, including welding procedure, weld geometry, and base metal dilution. Due to the complexity of the physical processes involved, FE analyses were employed to investigate the effects of these factors on residual stresses. Figure 4.6.1 shows four types of simulation models. Specifically, models for two different root penetration depths, a=1.7mm, b=1.7mm and a=5.0mm, b=1.7mm, are also considered to investigate their influence on residual stresses.

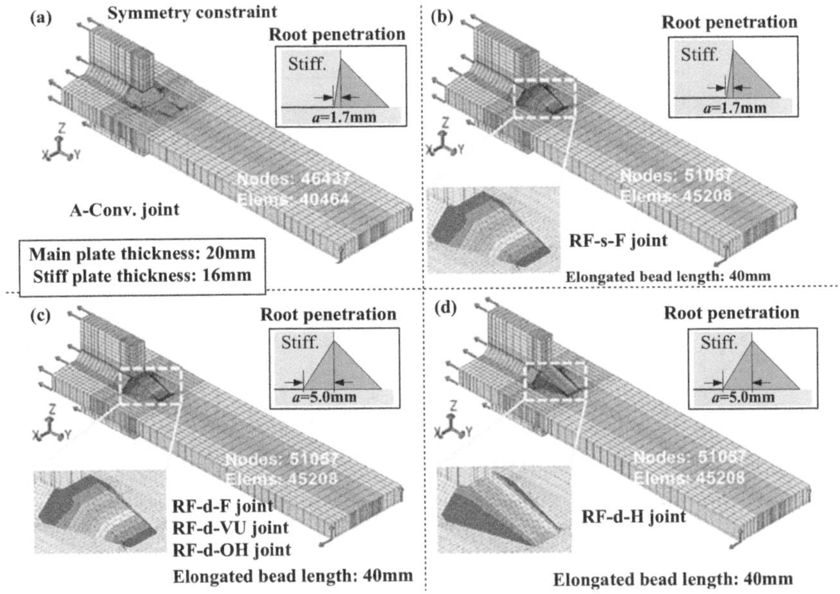

FIGURE 4.6.1
Simulation models of boxing fillet-welded joints.

To make the numerical analysis more efficient, a half-geometry of the welded joints was modeled by FE analysis, and a symmetric condition was applied along the left–right section. The arrows in Figure 4.6.1 represent boundary conditions that prevent rigid body motion of the model. The mesh size surrounding the elongated bead was approximately 2 mm, while a coarser mesh with a maximum size of 10 mm was employed for regions away from the elongated bead.

The conventional von Mises elastoplastic flow theory was taken into account:

$$d\varepsilon = d\varepsilon^{e} + d\varepsilon^{p} + d\varepsilon^{th} + d\varepsilon^{tr} \tag{4.6.1}$$

where ε^{e}, ε^{p}, ε^{th}, and ε^{tr} represent the increments of elastic strain, plastic strain, thermal strain, and transformation strain, respectively. Additionally, the isotropic hardening model was applied to residual stress computation.

4.6.2 Material Properties in FE Analyses

4.6.2.1 Dilution Rate of Elongated Bead in Different Welding Positions

The dilution rate of LTT weld metal can be estimated by the hardness and nickel content of the LTT elongated bead. Figure 4.6.2 shows the hardness and dilution rate distribution in the LTT elongated bead using 16%Cr–8%Ni LTT wire.

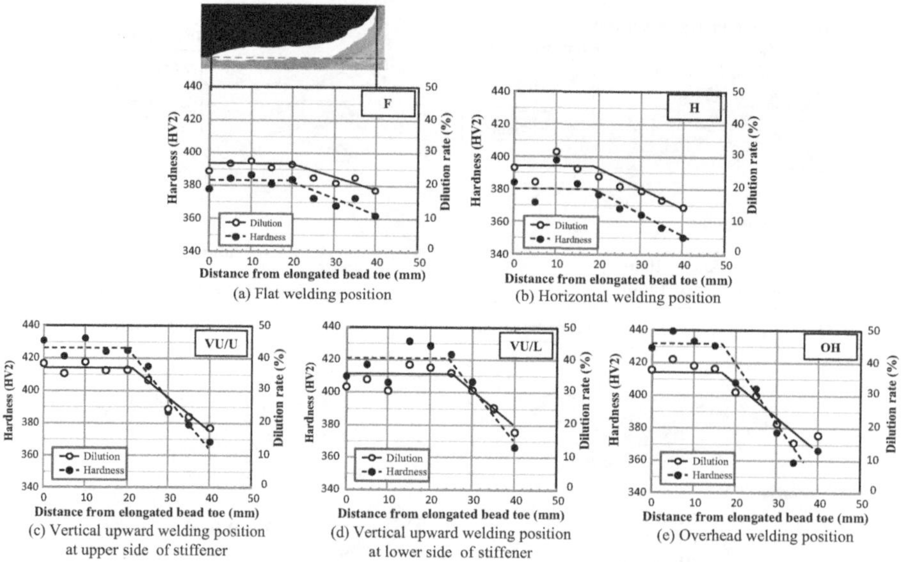

FIGURE 4.6.2
Dilution rate distribution estimated from the hardness distribution of the elongated bead in each welding position.

The elongated beads made in the flat and horizontal welding positions have a similar hardness of about 380–385 HV2 near the toe of the main plate side, while the hardness of the LTT elongated bead near the toe made in other welding positions is approximately 420–430 HV2. The Ni contents of the former and latter elongated beads are 6.1% and 5.2%, respectively. Correspondingly, the LTT elongated beads welded in the flat and horizontal positions have the same average dilution rate of 27%, while those welded in other positions have an identical dilution rate of 37% [25]. The difference in dilution rates is mainly due to the welding parameters and intermittent welding process.

Additionally, the hardness of the weld metal near the stiffener side in all welding positions was found to be around the 360 HV level, with a dilution rate of approximately 15%.

To simplify the numerical analysis, material properties corresponding to a dilution rate of 27% were used for the weld metal in flat and horizontal welding positions, while a dilution rate of 37% was applied for the weld metal in vertical-up and overhead welding positions.

4.6.2.2 Material Properties and Temperature Dependency

Based on the experimental measurements and physical assumptions, the thermal conductivity λ, specific heat c, and Young's modulus E for the base metal and weld metals, along with their temperature dependency, are shown in Figure 4.6.3. The measured thermal expansion curves during heating and cooling for the base metal and LTT welded metals are shown in Figure 4.6.4. The austenitization temperature for the base metal ranges from 700°C to 800°C, while for LTT welded metals, it ranges from 600°C to 700°C. From Figure 4.6.4, the martensite start (Ms) temperature can be estimated to be 200°C and 250°C, respectively, for the weld metal with dilution rates of 27% and 37%.

Figure 4.6.5 shows the yield strength (corresponding to the 0.2% strain in measured stress–strain curves obtained from uniaxial tensile tests) of base plate KA36, conventional weld metal, and LTT elongated bead, as well as their temperature dependency. The yield strength of welded metal with dilutions of 27% and 37% is 921 and 975 MPa, respectively, which is favorable for controlling variations in mechanical properties. In addition to thermal expansion, yield strength changes, and its temperature dependence during heating and cooling in the LTT welded metal, transformation-induced plasticity (TRIP) may also influence the magnitude of compressive residual stresses. For simplicity, it was not considered in this research because of the lack of measured data. Additionally, isotropic strain hardening was employed, and an annealing temperature of 800°C was defined in the numerical analysis. The density and Poisson's ratio of the three materials are assumed to be independent of temperature, with values of 7.85 g/cm^3 and 0.3, respectively.

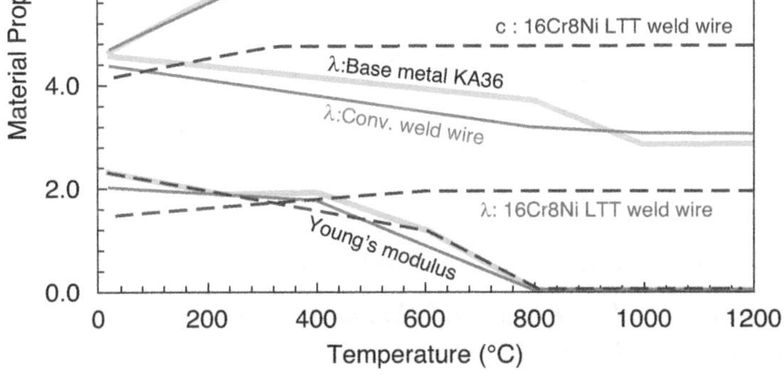

FIGURE 4.6.3
Temperature-dependent specific heat, thermal conductivity, and Young's modulus of KA36 base metal, DW-100 weld metal, and DW-168PF weld metal.

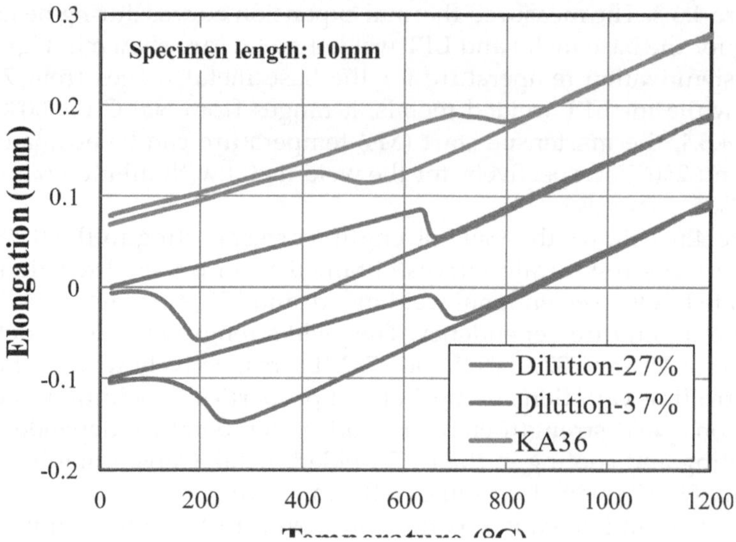

FIGURE 4.6.4
Measured thermal expansion of LTT elongated bead at dilution rates of 27% and 37%.

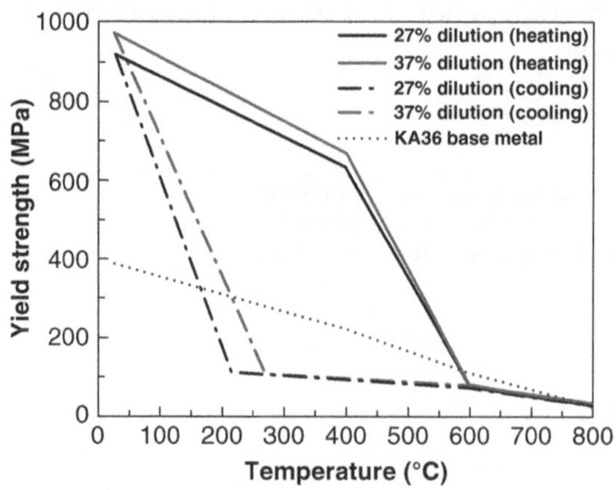

FIGURE 4.6.5
Temperature dependent yield strength of base metal KA36 and LTT weld metal with 27% and 37% dilution.

4.6.3 Numerical Analysis of Residual Stress

4.6.3.1 Thermal Cycles Due to LTT Elongated Bead Welding in Different Positions

In multi-pass welding, the heat input from the later weld bead is often insufficient to reheat the prior welds up to the austenitization temperature or to maintain the temperature above the Ms temperature, resulting in an accumulation of thermally induced strain and an increase in residual stress [26,27]. If the interpass temperature is controlled to be higher than the Ms temperature, it can be estimated that large compressive residual stresses will be generated throughout the multi-pass welds because the martensitic transformation occurs as a single entity.

For the LTT elongated beads, it is recommended to maintain the interpass temperature above the Ms temperature at the front weld toe. Considering the welding direction, it is critical to investigate the temperature distribution of elongated weld toes in the joints RF-d-H and RF-d-VU/L because their welding time is longer than other welding positions.

The temperature distribution of weld joint RF-d-H shown in Figure 4.6.6a reveals that all weld passes have interpass temperatures higher than Ms (220°C) except for the first weld pass, whose interpass temperature is between Ms and Mf. Fortunately, the reheat temperature in the first pass is higher than the Ac3 temperature of 700°C, allowing the martensitic transformation to occur together with the later welded metals.

The welding direction for the fillet weld in the horizontal position changes from the weld toe of the main plate side to the stiffener side, as shown in Figure 4.5.3b.

Multiple thermal cycles at all points 1–9 due to weaving welding exceed Ms (290°C) in the joint RF-d-VU/L, as shown in Figure 4.6.6b.

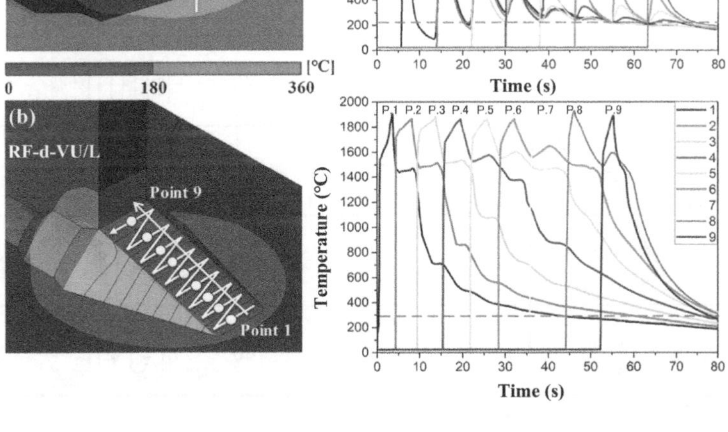

FIGURE 4.6.6
Thermal cycles in elongated bead welding: (a) RF-d-H and (b) RF-d-VU/L welded joints.

4.6.3.2 Numerically Analyzed Residual Stresses

Figure 4.6.7 summarizes six simulation cases that consider various welding conditions, including root penetration, dilution rate, and welding direction. Notably, the effect of dilution rate between the welding wire and base metal on material properties must be considered, as the chemical compositions of the resulting material differ from those of the base metal and weld wire.

Figure 4.6.8 shows the mid-section distribution of longitudinal residual stresses σ_y in all-position boxing fillet-welded joints with LTT elongated beads. In the conventional boxing fillet-welded joint without the LTT elongated bead (A-Conv-F), tensile residual stresses are induced, and a larger value exists at the front boxing bead toe, which is one of the main reasons why fatigue cracks often occur at the boxing weld toe. For boxing fillet-welded joints with LTT elongated beads, larger compressive residual stresses σ_y are produced in the LTT elongated bead, balanced by a large tensile residual stress in the base plate. Due to various welding conditions, there are slight differences in residual stress between elongated beads welded in different positions. Much larger compressive residual stresses are generated in the LTT elongated beads of boxing fillet joints RF-d-H, RF-d-VU/U, and RF-d-VU/L. Furthermore, it can also be seen that tensile residual stress is induced at the root of the boxing fillet-welded joint. The tensile residual stress at the root of specimen RF-s-F is larger than those of specimens RF-d-F, RF-d-H, RF-d-VU/U, and RF-d-VU/L, which is consistent with the result in Figure 4.4.5. This means that deep root penetration reduces the tensile residual stresses at the root.

Simulation cases	Welding position	Penetration *a* (mm)	Dilution rate (%)	Elongated bead length (mm)	Welding direction	Simulation model
A-Conv.	Flat	1.7	-	-		Model (a)
RF-s-F	Flat	1.7	27	40	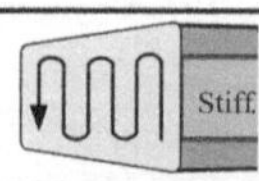	Model (b)
RF-d-F	Flat	5	27	40	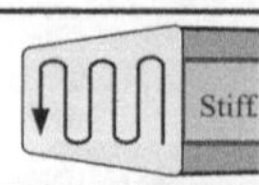	Model (c)
RF-d-H	Horizontal	5	27	40	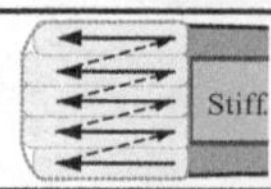	Model (d)
RF-d-VU/U	Vertical upward (Upper side)	5	37	40	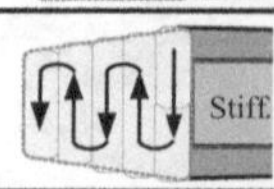	Model (c)
RF-d-VU/L	Vertical upward (Lower side)	5	37	40	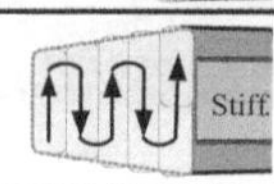	Model (c)

FIGURE 4.6.7
Simulation cases for residual stress computation.

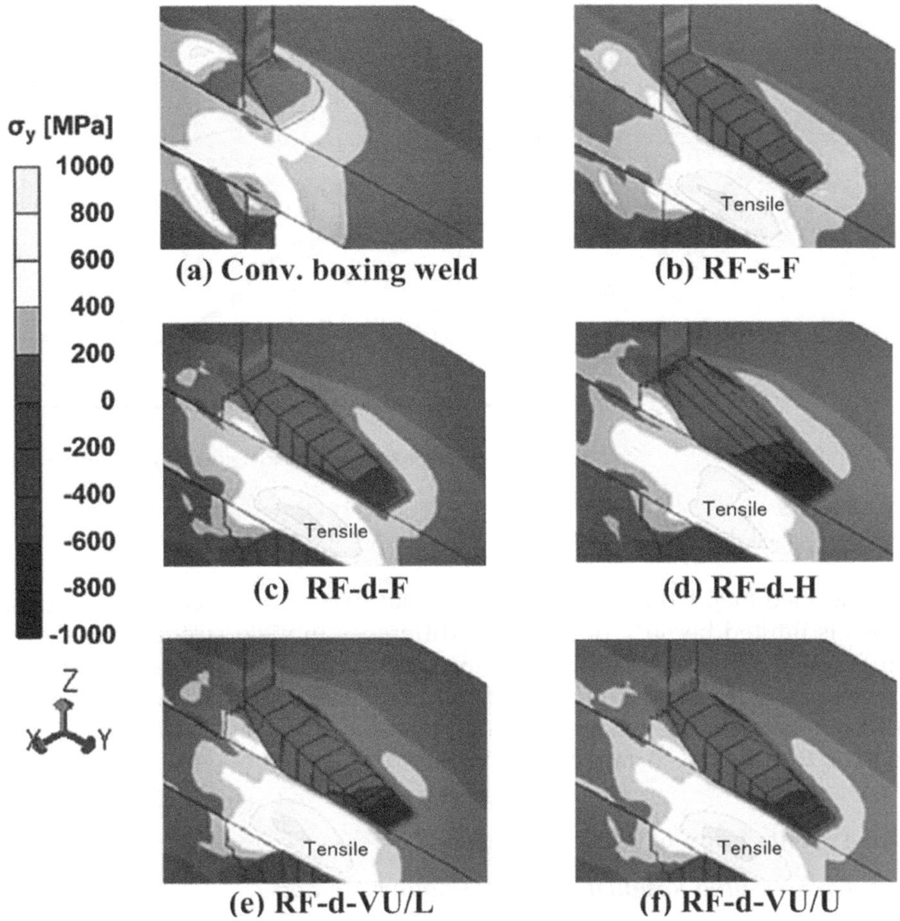

FIGURE 4.6.8
Residual stress σ_y of boxing fillet-welded joints (mid-section).

Figure 4.6.9 shows the comparison among computed longitudinal residual stress distributions along the line *L* from the weld toe on the stiffener side to the weld toe on the base plate side. The conventional boxing fillet-welded joint and that with an elongated bead welded using conventional welding wire, as expected, have higher tensile residual stresses at the boxing weld toe, with values reaching up to 350 and 500 MPa. The residual stress distribution is similar among LTT elongated beads welded in different positions. The compressive residual stresses on the LTT elongated bead surface are between −200 and −600 MPa, with a maximum value reaching −800 MPa near the weld toe. The longitudinal compressive residual stress on the LTT elongated bead due to the horizontal welding position is larger than in other welding positions. This is because the welding direction is absolutely the same as to the direction of longitudinal stress σ_y. Compared with the yield strength of

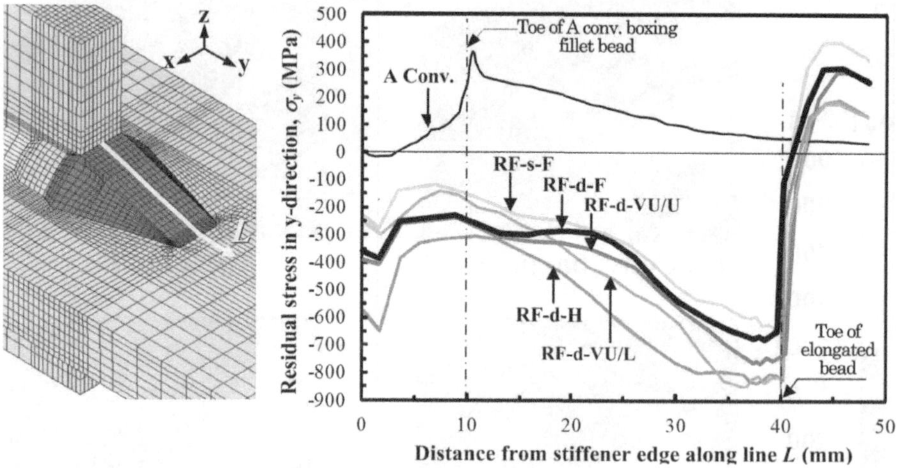

FIGURE 4.6.9
Surface residual stress distribution along the LTT elongated bead welded in various positions.

LTT welded metal, the longitudinal residual stress difference across all positions is limited because of the small difference in yield strength, as shown in Figure 4.6.5, which is favorable for practical applications and meets our development purposes. These compressive residual stresses existed in the LTT elongated beads are large enough to reduce the risk of stress concentration on the uneven surface of the weld bead.

4.6.4 Verification of Analyzed Residual Stress

4.6.4.1 Measured Residual Stress by Contour Method

The residual stress distribution in the elongated bead of the repaired RP-d-F type specimen was measured before the fatigue test using the contour method [28,29]. Figures 4.6.10 and 4.6.11 show the residual stress distribution on the cutting surface. It can be seen that large compressive residual stress of up to −680 MPa can be generated in the LTT elongated beads. At the center of plate thickness, tensile residual stress is present.

Generally, the compressive residual stress in the LTT elongated bead is dominated by the martensitic transformation that occurred in the LTT weld metals. At the same time, tensile residual stress exists in the vicinity of the LTT welded bead to maintain equilibrium. Additionally, as the base plate thickness is 20 mm, the influence of the welding sequence on the residual stress on side 2 and side 4 (refer to Figure 4.3.1) is not strong. Therefore, only a minor difference in compressive residual stress can be observed in the elongated bead between side 2 and side 4. Upon closer examination, the tensile residual stress in the base plate near side 4 has a larger value because LTT elongated bead 4 was welded later.

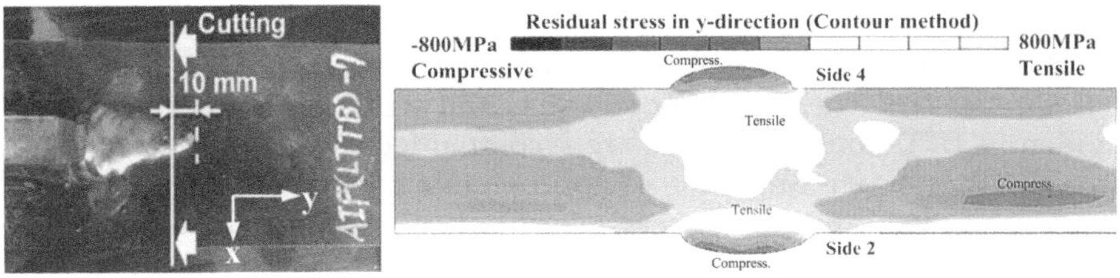

FIGURE 4.6.10
Longitudinal residual stress at the section 10 mm from the weld toe, measured by the contour method.

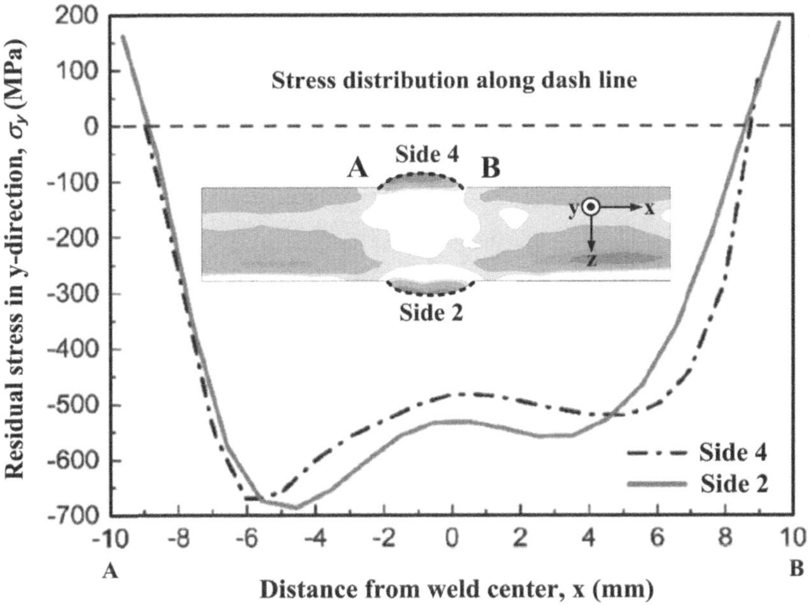

FIGURE 4.6.11
Longitudinal residual stress distribution along the dashed line AB.

4.6.4.2 Measured Residual Stress by X-Ray Diffraction Method

Former studies by Barsoum and Gustafsson [30,31] reported the effect of load conditions on stress relaxation during fatigue testing using a martensitic welding wire. Based on their studies, the residual stresses remained stable when applying constant low amplitudes in comparison with varying loads. As a result, greater fatigue enhancement was achieved for specimens tested under constant amplitudes. In the current study, it is interesting and meaningful to investigate the final stress state of the LTT elongated bead after the fatigue test.

Figure 4.6.12 shows the measured longitudinal residual stress in the elongated beads of the boxing fillet-welded joints RF-d-VU/L after the fatigue test using X-ray diffractometer. The fatigue test conditions were as follows: air atmosphere at room temperature, cycle speed 3–5 Hz, stress ratio $R=0.0$, sine wave as the load wave, and stress ranges $\Delta\sigma$ of 150 MPa. Evidently, LTT elongated beads are accompanied by compressive residual stress with a maximum value of approximately −643 MPa, which has the same level of the residual stress measured by the contour method. This means that the stress relaxation during the fatigue test is minimal in the LTT elongated bead.

Here, the numerical analysis results of the RF-d-VU/L specimen (refer to Figure 4.6.9) and the actual measurement results (■, □) obtained by X-ray diffractometer were compared, as shown in Figure 4.6.13.

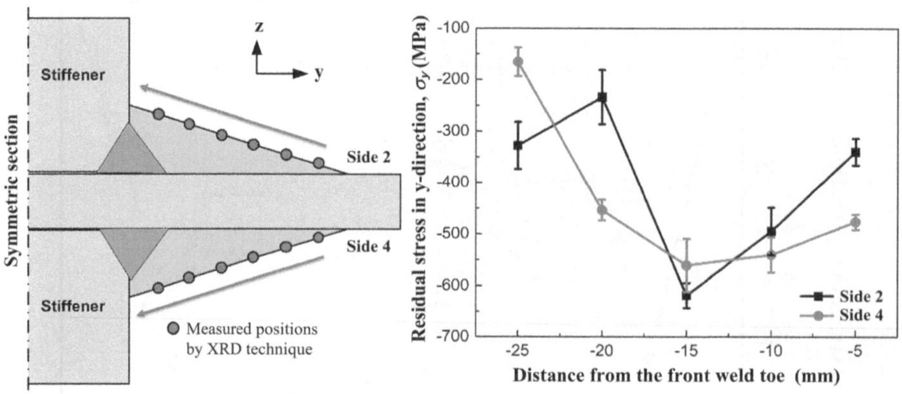

FIGURE 4.6.12
Longitudinal residual stress measured after the fatigue test.

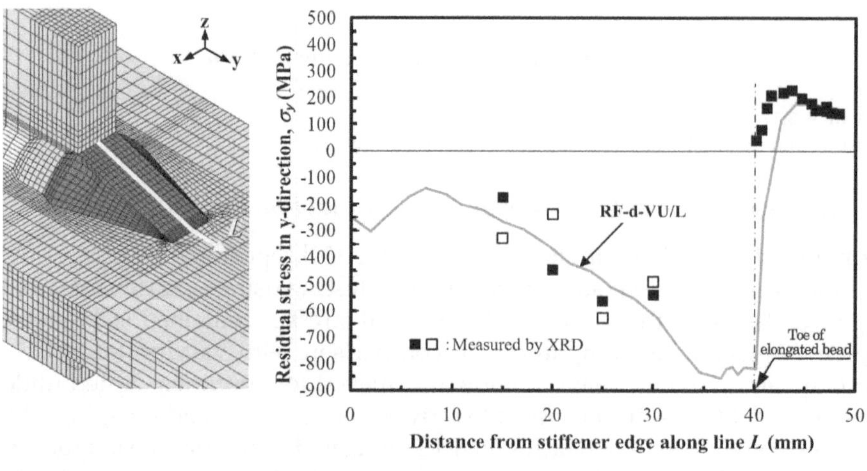

FIGURE 4.6.13
Comparison with measured residual stress in the RF-d-VU/L specimen.

The measured residual stress on the elongated bead surface is between −200 and −600 MPa. Both the measured and computed residual stresses at a position 5 mm away from the weld toe are tensile, with a range at the +200 MPa level.

From these results, it can be concluded that there is generally good agreement, demonstrating the validity of the numerical analysis.

4.7 Development of LTT Elongated Bead Welding Technology for Fatigue Strength Enhancement through Repair and Reinforcement

In semi-automatic welding at ship repair sites, various welding defects often occur. To mitigate these defects, Matsuzaki et al. [23] compiled a practical technical know-how into the "Welding Construction Guidelines" (refer Sections 4.5.2 and 4.5.3). The following are the fatigue test results of repair and reinforced elongated bead weld joint specimens manufactured under each welding position utilizing this technical know-how.

Additionally, the fatigue test was terminated when the number of fatigue cycles reached approximately six times the mean fatigue life $[N_f]_{\text{Conv.mean}}$ of the conventional boxing fillet-welded joint.

4.7.1 Fatigue Test Results of Welded Joints in Each Welding Position and Optimization of Welding Procedure

The fatigue test results of the specimens constructed in each welding position are summarized in Table 4.7.1.

All specimens fabricated in the flat welding position did not fracture until approximately six times $[N_f]_{\text{Conv.mean}}$. Compared to the fatigue test results in the flat position discussed in Section 4.3, the root reinforced by $a=5$ mm and $b=3$ mm is extremely effective.

However, for intermittent welding in other positions, as noted in the Remarks column of Table 7.4.1, welding defects occurred, resulting in a shorter fatigue life extension than six times. Particularly in joints welded in the overhead position, it was difficult to completely prevent welding defects. Representative examples of these defects are shown in Figure 4.7.1. The defect in Figure 4.7.1a shows slag inclusion in the repair bead, Figure 4.7.1b shows surface irregularities at the gusset edge caused by slag inclusion in the repair bead, Figure 4.7.1c shows undercut at the gusset edge in the continuous bead, and Figure 4.7.1d illustrates lack of fusion with residual material during gouging of the repair bead. The figure shows the fracture surfaces and bead appearances of test specimens containing these defects.

TABLE 4.7.1
Fatigue Test Results of Welded Joints Under Each Welding Position

Welding Position	No.	Specimen Type	Stress Range $\Delta\sigma$ (MPa)	Cycles to Failure N_f (Cycles)	Ratio of Conv. Mean	Remarks
Flat	1	RF-16Cr8Ni-F	150	>2,032,789	>6.1	Run out
	2	RP-16Cr8Ni-F	120	>3,540,094	>5.4	Run out
	3	RP-16Cr8Ni-F	120	>4,300,000	>6.6	Run out
	4	RP-16Cr8Ni-F	150	>2,100,000	>6.3	Run out
	5	RP-16Cr8Ni-F	175	>1,500,001	>7.1	Run out
	6	RP-16Cr8Ni-F	200	>961,356	>6.8	Run out
Horizontal	1	RF-16Cr8Ni-H	200	338,739	2.4	Slag inclusion
	2	RF-16Cr8Ni-H	150	>2,070,000	>6.2	Run out
	3	RF-16Cr8Ni-H	120	>4,000,000	>6.1	Run out
	4	RP-16Cr8Ni-H	200	>850,000	>6.0	Run out
	5	RP-16Cr8Ni-H	150	>2,000,001	>6.0	Run out
	6	RP-16Cr8Ni-H	120	>4,120,000	>6.9	Run out
Vertical upward at the upper side of stiff	1	RF-16Cr8Ni-VU/U	150	>2,100,000	>6.3	Run out
	2	RF-16Cr8Ni-VU/U	120	2,876,048	4.4	Undercut at stiff. end
	3	RF-16Cr8Ni-VU/U	200	426,093	3.0	Undercut at stiff. end
	4	RP-16Cr8Ni-VU/U	150	>2,000,000	>6.0	Run out
	5	RP-16Cr8Ni-VU/U	120	>4,000,000	>6.1	Run out
	6	RF-16Cr8Ni-VU/U	200	533,563	3.8	Lack of fusion
	7	RF-16Cr8Ni-VU/U	200	461,098	3.3	Surface of elongated bead
Vertical upward at the lower side of stiff	1	RP-16Cr8Ni-VU/L	200	772,758	5.5	Undercut at stiff. end
	2	RP-16Cr8Ni-VU/L	120	1,656,754	2.5	Undercut at stiff. end
	3	RP-16Cr8Ni-VU/L	120	4,000,000	>6.1	Run out
	4	RF-16Cr8Ni-VU/L	150	1,400,252	4.2	Lack of fusion
	5	RF-16Cr8Ni-VU/L	150	2,000,000	>6.0	Run out
	6	RF-16Cr8Ni-VU/L	200	361,036	2.6	Lack of fusion
	7	RF-16Cr8Ni-VU/L	200	513,361	3.7	Surface of elongated bead
Overhead	1	RF-16Cr8Ni-OH	150	811,820	2.4	Undercut at stiff. end
	2	RP-16Cr8Ni-OH	140	659,888	2.2	Slag inclusion
	3	RP-16Cr8Ni-OH	150	1,347,744	4.0	Slag inclusion
	4	RP-16Cr8Ni-OH	120	1,049,754	1.6	Slag inclusion
	5	RP-16Cr8Ni-OH	200	>850,000	>6.0	Run out
	6	RP-16Cr8Ni-OH	120	>4,000,000	>6.1	Run out
	7	RF-16Cr8Ni-OH	150	994,875	3.0	Slag inclusion
	8	RF-16Cr8Ni-OH	150	604,831	1.8	Slag inclusion
	9	RF-16Cr8Ni-OH	150	>2,000,000	>6.0	Run out

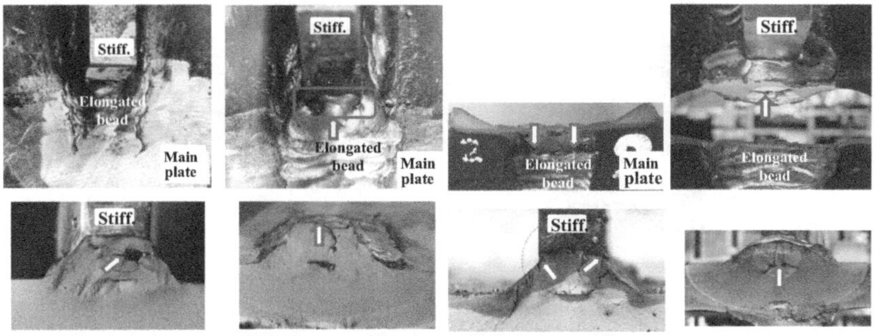

(a) Slag inclusion (RP-16Cr8Ni-OH_4) (b) Undercut at stiff. end (RF-16Cr8Ni-OH_1) (c) Undercut at stiff. end (RF-16Cr8Ni-VU/U_3) (d) Lack of fusion (RF-16Cr8Ni-VU/L_6)

FIGURE 4.7.1
Examples of welding defects in re-boxing fillet bead and elongated bead.

The following improvement points have been presented:

1. Slag removal was thoroughly implemented after each pass during re-boxing fillet welding.
2. Not only for the elongated bead welding but also considering the dependency on the bead shape at the edge of the gusset plate during re-boxing fillet welding, the torch angle and bead manipulation rules were established. Even if shape irregularities occur, grinding for dressing was permitted.
3. For the surfaces to be welded, not only slag but also spatter and residues from gouging, etc., must be removed.

These procedures were documented in the "Welding Construction Guidelines".

Except for the fatigue results of joints in the overhead welding position and those in the vertical upward welding position under a stress range of $\Delta\sigma$=200 MPa, these process improvements have ensured a six-fold fatigue life extension $[N_f]_{\text{Conv.mean}}$ in the test results.

From these fatigue results, the following conclusions can be drawn:

1. It was thought that the harmful effects of the repair weld could be mitigated, as neither the repair specimen nor the reinforced specimen fractured, and the crack repair weld was covered by the LTT elongated bead.
2. These results likely indicate that the stress concentration caused by the uneven surface of the elongated bead in intermittent welding is offset by the high compressive residual stress on its surface.

However, in the high stress range of $\Delta\sigma$=200 MPa, as shown in Figure 4.7.2, no internal defects such as slag inclusions were observed on the fracture

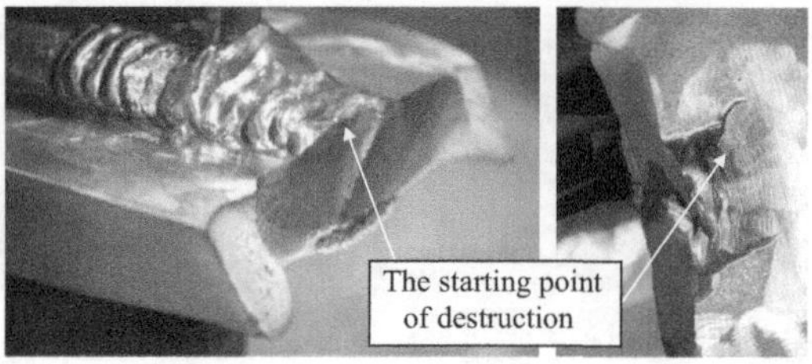

FIGURE 4.7.2
Fracture situation of the welded joint in a vertical upward position at $\Delta\sigma$=200 MPa.

surface of the joint in the vertical upward welding position. Instead, failure originated from the uneven elongated bead surface, suggesting that the compressive residual stress could no longer compensate for the significant stress concentration at the intermittent bead surface.

Then, the improvement of the welding procedure in the overhead welding position is discussed.

For the test specimen that exhibited bead irregularities, as shown in Figure 4.7.1b, improvement point (2) was followed: the LTT elongated bead was completely removed with a grinder, and then the elongation bead welding was performed again. The resulting test specimens are RP-16Cr8Ni-OH_5 and -OH_6. Both test specimens achieved a long fatigue life of six times of $[N_f]_{\text{Conv.mean}}$.

On the other hand, regarding the internal defects in Figures 4.7.1a and 4.7.3a, the results of various attempted countermeasures revealed that they consistently occur at the junction between the 3rd and 4th pass bead welding start points within the dashed circle in Figure 4.7.3b.

Therefore, the welding process was changed to a re-boxing fillet welding method that eliminates the junction at the welding start point. The modification procedure is shown in Figure 4.7.4.

1. Gouging process to ensure penetration parameter *a* at the stiffener end
2. Re-welding work at the stiffener end
3. Gouging process to ensure penetration parameter *b* at the stiffener sides
4. Re-welding work at the fillet part of the stiffener sides

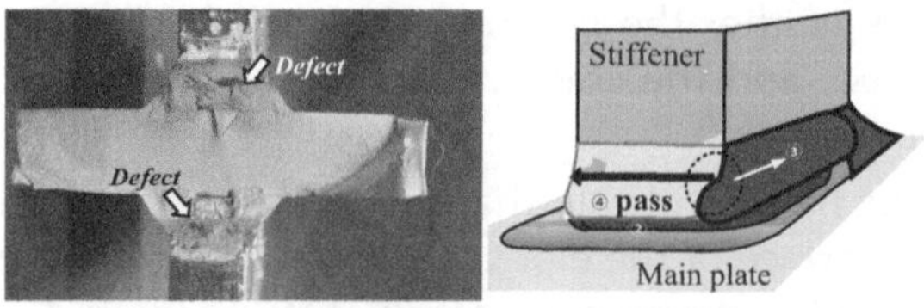

(a) RP-16Cr8Ni-OH8 (b) Schematic view

FIGURE 4.7.3
Internal defects of re-boxing fillet weld in the overhead welding position.

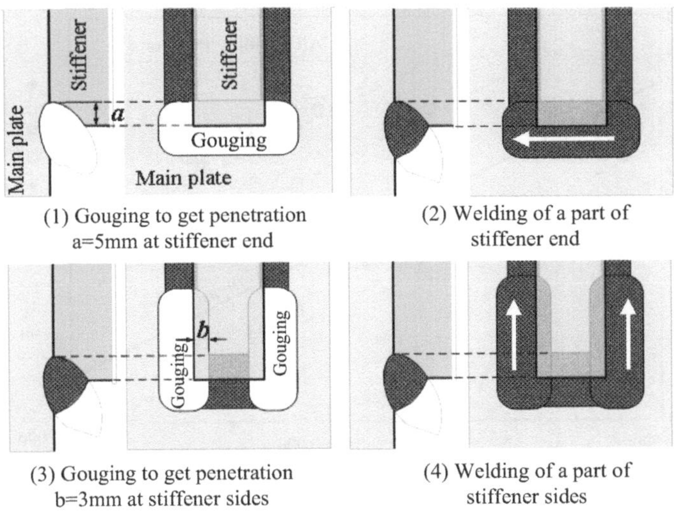

FIGURE 4.7.4
Improved re-boxing fillet weld method for the overhead welding position.

The fatigue test results of the specimens (RF-16Cr8Ni-OH_9) fabricated using the new welding procedure confirmed that the fatigue life could be extended by more than six times, verifying the effectiveness of this new procedure.

This procedure is also considered useful for construction in the overhead welding position and is expected to enhance the reproducibility of defect-free construction using the semi-automatic arc welding method.

4.7.2 Summary

For practical use in ship repair sites, the LTT elongated bead welding repair method, applicable to all welding positions, has been developed using semi-automatic arc welding.

As previously described, the use of FCWs during the development phase led to welding defects such as lack of fusion and slag inclusions, which hindered the improvement of fatigue strength. However, through repeated improvements in the welding procedures, a significant increase in the fatigue life of welded joints constructed in various welding positions was ultimately achieved, as shown in Figure 4.7.5, and the "Welding Construction Guidelines" were established.

In particular, for repair welding using the LTT elongated bead welding method, the effect of repair welding after crack removal does not impact fatigue strength, making it an extremely simple and practical repair method.

Furthermore, the fatigue test results conducted in various welding positions (Figure 4.7.5) demonstrate that fatigue life is guaranteed to extend by more than six times. However, limitations in fatigue life extension were

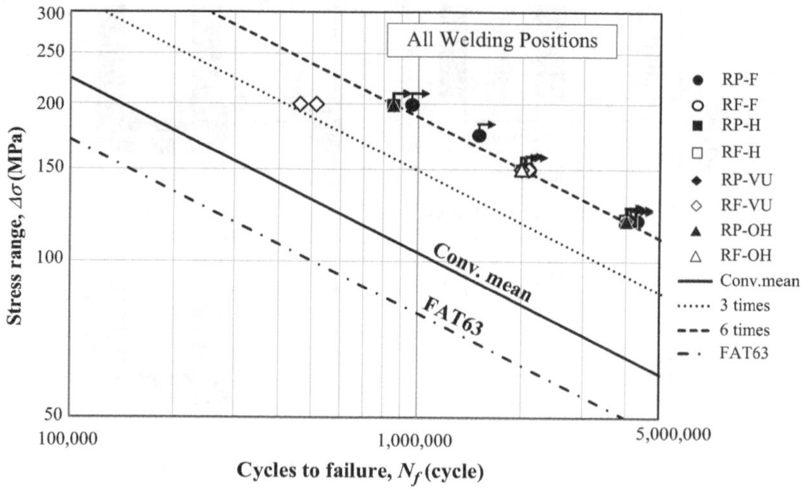

FIGURE 4.7.5
Relationship between the stress range and the number of repeated cycles to failure without welding defects.

observed only in the high-stress range ($\Delta\sigma$=200 MPa) of the vertical upward welding position; nevertheless, an extension of approximately four times was ensured.

The above results are from fatigue tests conducted on test specimens welded at ship repair sites. This welding method is deemed suitable for practical use and has obtained welding procedure approval from the Classification Society (Class NK).

4.8 Numerical Analysis for Application of LTT Elongated Bead Method to Fillet-Welded Joints with Different Thickness Combinations

As mentioned in Section 4.7, fatigue tests on large-scale specimens reinforced and repaired with the LTT elongated bead welding method in various welding positions confirmed significant fatigue life extension under multiple stress ranges [21,23]. However, fatigue experiments were only conducted on one combination of main plate thickness (t_p=20 mm) and stiffener plate thickness (t_s = 16 mm). To extend this method to practical structures, it is necessary to verify its effectiveness on different plate thickness combinations.

In this section, numerical analysis using the FEM program *JWRIAN* developed at Osaka University [24,32–36] was conducted to evaluate the stress concentration factor, compressive residual stress distribution, and stress intensity factor at the root for various plate thickness combinations of boxing fillet-welded joints using the LTT elongated bead with designed size.

4.8.1 Numerical Analysis on Reference Model of Fatigue Test

4.8.1.1 Fatigue Test Specimen and Reference Model

Figure 4.8.1 shows the LTT elongated bead welded joint test specimen (referred to as the reference fatigue test specimen) with a main plate thickness of $t_p = 20$ mm, stiffener plate thickness of $t_s = 16$ mm, and length of 800 mm.

Figure 4.8.2a is a macro photograph of the longitudinal section of a weld zone where elongated bead welding was performed after boxing fillet welding. This cross-section is modeled in Figure 4.8.2b.

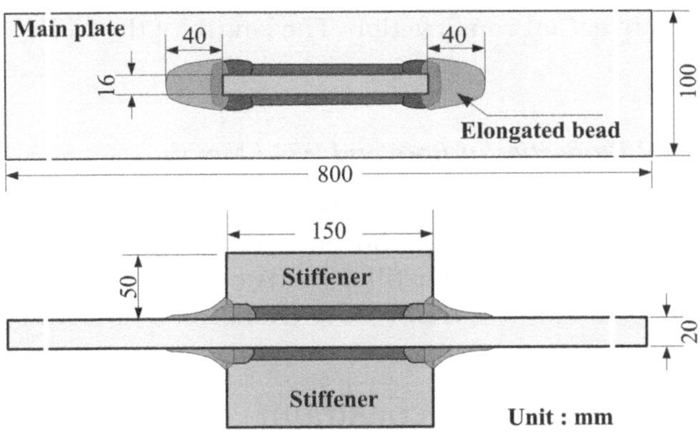

FIGURE 4.8.1
Experimental fatigue test piece of boxing fillet-welded joint with LTT elongated bead.

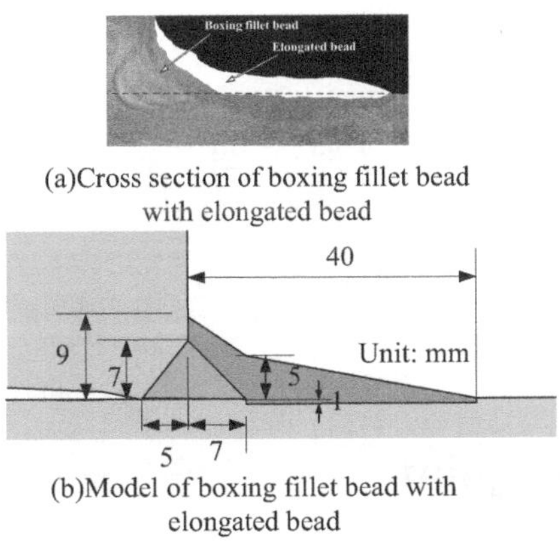

FIGURE 4.8.2
Model and dimensions of the boxing fillet bead with an elongated bead in plate thickness (t_p)/stiffener thickness (t_s) = 20 mm/16 mm.

Figure 4.8.3 presents the finite element model for conventional boxing fillet-welded joints, while Figure 4.8.4 shows the analysis model (referred to as the baseline model) for LTT elongated bead joints, where the LTT elongated bead is deposited over the conventional boxing fillet weld, as shown in Figure 4.8.3. To reduce computation time, only half of the joint length was modeled, taking advantage of symmetry. The arrows in the figures indicate the directions of constrained nodal displacements for symmetry and rigid body motion. The element sizes ranged from 0.2 to 1.0 mm in the weld metal and elongated bead toe regions and from 0.2 to 10 mm in the base metal.

For conventional boxing fillet welds, the penetration depths at the stiffener end and side were set to 1.7 mm (Figure 4.8.3). For the LTT elongated bead joint model (Figure 4.8.4), based on previous results [23], the penetration depths *a* and *b* were set to 5 and 3 mm, respectively, achieved through air arc gouging in actual construction. The length of the LTT elongated bead was 40 mm.

4.8.1.2 Material Properties of Base and Weld Metals

The base metal for the main plate and stiffener was shipbuilding high strength steel KA36. For welding, conventional FCW (Kobelco: DW-100, hereafter "Conv. wire") was used for fillet welds on the stiffener side, while the newly developed 16%Cr–8%Ni LTT FCW (Kobelco: DW-168FP) [23] was used

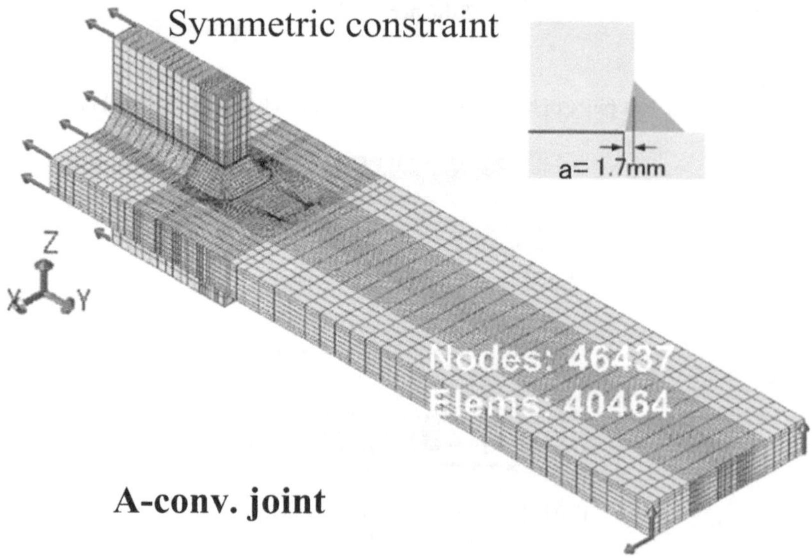

FIGURE 4.8.3
Finite element mesh of the conventional box welded joint.

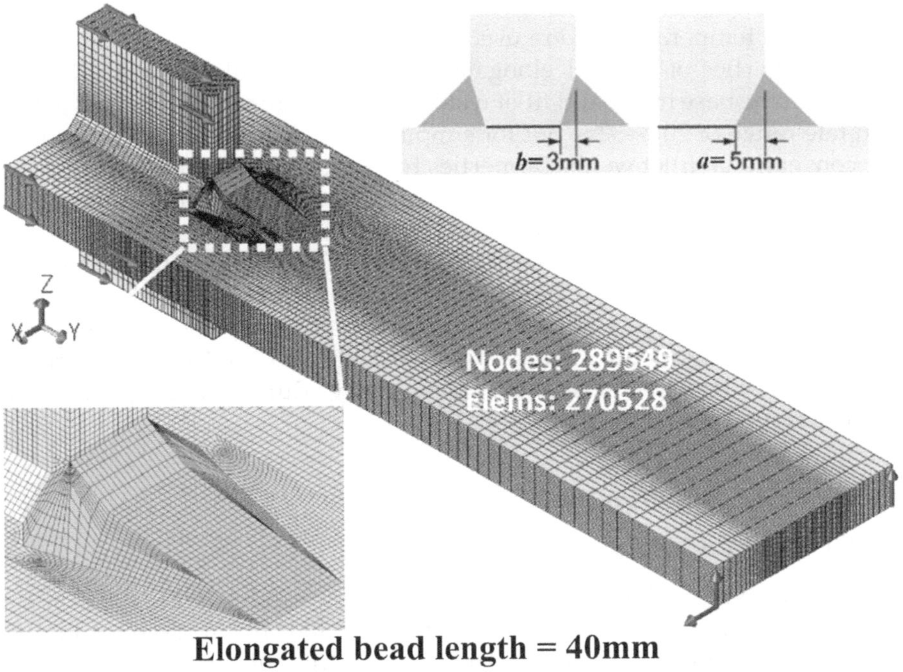

FIGURE 4.8.4
Mesh of box welded joint with LTT elongated bead.

for the elongated bead. Table 4.8.1 lists the chemical compositions of KA36, conventional welding wire, and LTT wire.

To analyze stress concentration, welding residual stress, and stress intensity factors for conventional and elongated bead joints, the material properties of the base and weld metals were required.

For stress concentration and stress intensity factor analysis, only Young's modulus and Poisson's ratio were needed. For residual stress analysis, properties such as density, specific heat, thermal conductivity, thermal expansion

TABLE 4.8.1

Chemical Compositions (%) of Base Metal KA36, Conventional Welding Wire (DW-100), and 16%Cr–8%Ni LTT Welding Wire (DW-168FP)

	Chemical Composition (Mass%)						
Material	**C**	**Si**	**Mn**	**P**	**S**	**Ni**	**Cr**
Steel KA36	0.16	0.24	1.04	0.01	0.005	0.01	0.01
16Cr8Ni LTT wire (DW-168FP)	0.02	0.34	0.44	0.02	0.01	8.2	16.1
Conv. wire[a] (DW-100)	0.05	0.45	1.35	0.013	0.009	–	–

[a] AWS A5.20E71T-1C.

coefficient, yield stress, and work-hardening coefficient from room temperature to high temperatures were used.

The properties of the LTT elongated bead weld metal vary with dilution rates from the base metal and fillet weld metal. Measurements showed a dilution rate range of 20%–32%, but for simplicity, the average of 27% was used. For conventional fillet welds, properties from the weld metal were applied.

Figures 4.6.2 and 4.6.5 in Section 4.6 present the temperature-dependent physical (specific heat, thermal conductivity) and mechanical (Young's modulus, yield stress, tensile strength) properties of the base metal, Conv. wire fillet weld metal, and LTT elongated bead weld metal (27% dilution). Above 800°C, properties were assumed to be constant. Density and Poisson's ratio were uniform across all metals at 8.0×10^{-6} kg/mm^3 and 0.3, respectively. Figure 4.8.5 shows the thermal expansion curves during cooling, indicating transformation characteristics.

4.8.2 Analysis Results of the Reference Model

4.8.2.1 Stress Concentration Analysis Results

For the conventional boxing fillet-welded joint model (Figure 4.8.3), the maximum principal stress distribution under a 100 MPa average load in the Y-direction is shown in Figure 4.8.6 [32]. The stress at the main plate weld toe A, stiffener weld toe B, and root R is 267, 166, and 159 MPa, respectively, indicating stress concentration factors of 2.67, 1.66, and 1.59.

For the elongated bead joint model (Figure 4.8.4), the stress concentration factors shown in Figure 4.8.7 [32] are 1.33, 1.45, and 1.07 at weld toes A, B, and root R, respectively. It is 1.65 at weld toe C, located at the geometric

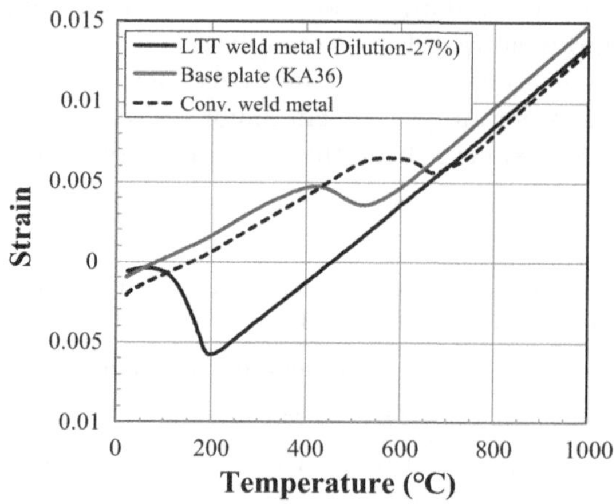

FIGURE 4.8.5
Measured thermal expansion during cooling of KA36 base metal, conv. weld metal, and LTT weld metal.

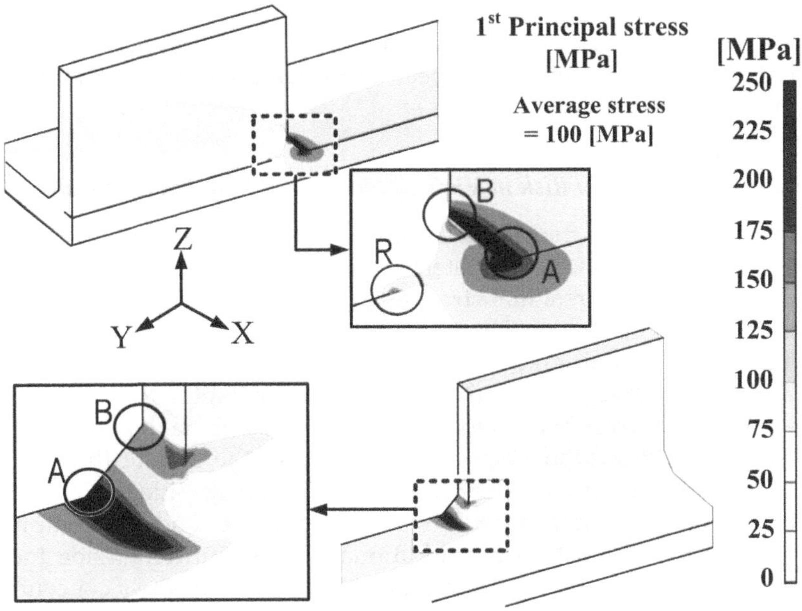

FIGURE 4.8.6
Analyzed stress concentration in the conventional boxing fillet-welded joint at an average stress of 100 MPa.

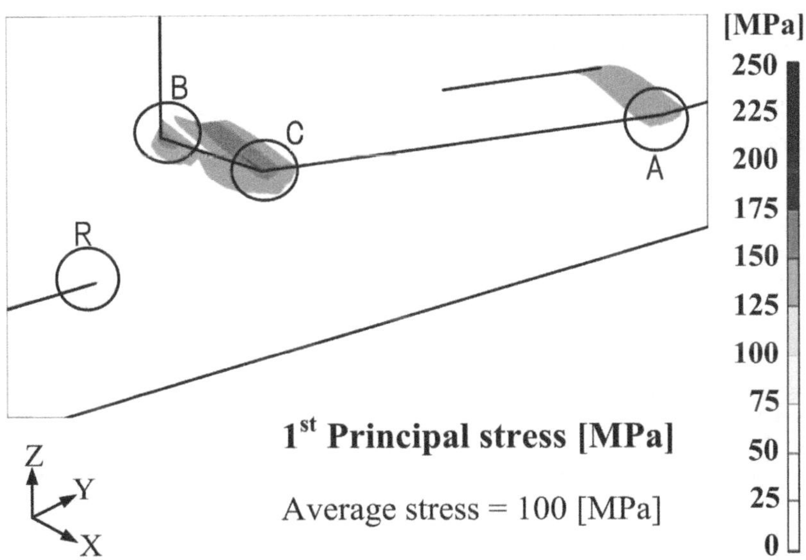

FIGURE 4.8.7
Analyzed stress concentration in the boxing fillet-welded joint with elongated bead at an average stress of 100 MPa.

change position of the elongated bead surface. Thus, stress concentration is reduced by 50% at weld toe A, 10% at toe B, and 30% at root R, demonstrating the effectiveness of the elongated bead in mitigating stress concentration at critical positions.

4.8.2.2 Crack Initiation Risk in View of Numerical Simulation Results

To qualitatively assess the risk of fatigue crack initiation, the stress concentration-induced stress (σ_F) generated at the weld toe ends under a 200 MPa load was added to the residual stress (σ_R) obtained from the aforementioned model's calculation results. This combined stress was used as the evaluation stress (σ_T) and compared between the conventional boxing fillet weld toe ends and the elongated weld toe ends, as shown in Table 4.8.2.

The evaluation stress at each toe end of the conventional boxing fillet weld resulted in significant tensile stress. However, in the case of the LTT elongated weld bead, the evaluation stress at the main plate-side toe end became compressive due to the reduced stress concentration and the compressive residual stress induced by LTT. Additionally, at the stiffener-side toe end, the stress level was sufficiently low, indicating that the stress concentration component was largely offset by the compressive residual stress.

Regarding these results, Figure 4.8.8 shows the numerical analysis results of the residual stress distribution at the root of the boxing fillet weld, presenting the distribution along the width direction (X) of the main plate, including the root. At the root, tensile residual stresses of 300 MPa in σ_z, and 150 MPa in σ_y and σ_x are generated.

At the root area, there is no beneficial effect from the compressive residual stress generated by the LTT elongated bead. Numerical analysis indicates that the root area has a high risk of fatigue crack initiation, which supports the findings of the fatigue test results.

TABLE 4.8.2

Result of Estimations on Evaluation Stress from Data

	A-Conv. Boxing Bead		LTT Elongated Bead	
	Toe at Main Plate	**Toe at Stiffener Plate**	**Toe at Main Plate**	**Toe at Stiffener Plate**
Stress concentration factor	2.67	1.59	1.33	1.45
Stress in 200 MPa load σ_F (MPa)	534	318	266	290
Residual stress σ_R (MPa)	262	2	–711	–481
Evaluation stress σ_T (MPa)	796	320	–445	–191

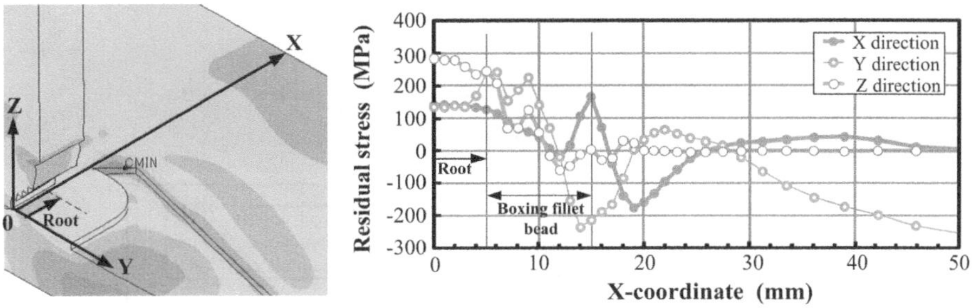

FIGURE 4.8.8
Analyzed distribution of residual stresses near the root in the boxing fillet-welded joint with elongated bead.

4.8.3 Numerical Analysis on LTT Elongated Bead Models with Different Thickness Combinations

Fatigue tests [21] revealed that cracks propagated from the root R and reached the stiffener weld toe, leading to complete failure. To assess the risk of root crack propagation, the root gap was treated as a crack whose length is equal to the root width (t_s–2b). The stress intensity factor (K) was computed using the characteristic tensor method [24].

Without consideration of residual stress, K decreases with deeper penetrations a and b as shown in Figure 4.8.9 [32]. For $a = 5$ mm and $b \approx 0$ mm, K has a peak value of 228 N/mm$^{3/2}$, correlating with a 3–4 times extension of fatigue life before root failure [21]. For the reference specimen (plate thickness: 20 mm, stiffener thickness: 16 mm) with penetrations ($a = 5$ mm, $b = 3$ mm), no root failure occurred even after six times the fatigue life extension [23].

4.8.3.1 Plate Thickness Combinations in Ship Structures

Most ship structural members (excluding extremely thick sections) have plate thicknesses of 6.0–30.0 mm. To evaluate optimal construction conditions, stress concentration, residual stress, and stress intensity factor K were analyzed for various t_p and t_s combinations (Table 4.8.3). Penetration depths (a, b), elongated bead width (W_{bead}), and leg length (H_{bead}) were specified, with a constant bead length of 40 mm.

4.8.3.2 Stress Concentration Evaluation

Table 4.8.4 lists stress concentration factors for critical locations (A, B, C, R) across thickness combinations. The relationship among stress concentration factors and two thickness combinations (t_p, t_s) can be expressed by a function in Eq. (4.8.1) and the response surface shown in Figure 4.8.10.

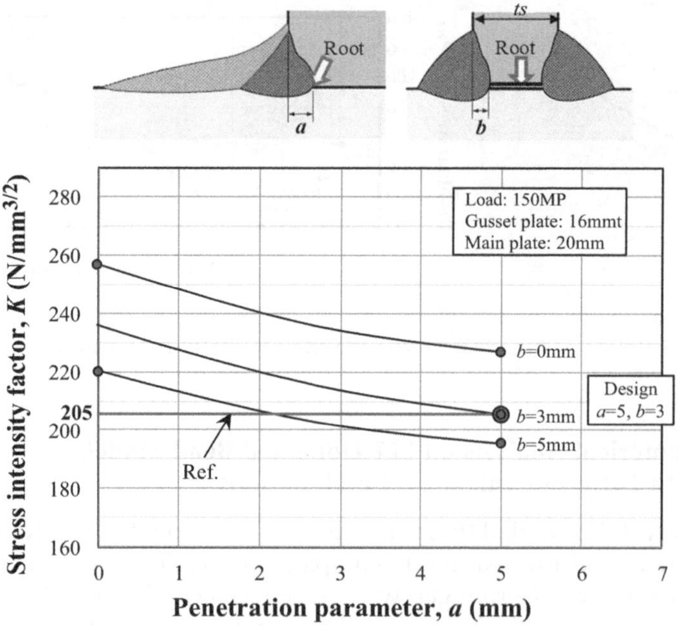

FIGURE 4.8.9
Influence of penetration parameters *a* and *b* on the stress intensity factor *K* at an average stress of 150 MPa.

TABLE 4.8.3
FE Analyzing Cases for Various Thicknesses of Main Plate and Stiffener

Model Name	t_p mm	t_s mm	a mm	b mm	W_{bead} mm	H_{bead} mm
p20/s16-RS	20	16	5	3	34	9
p20/s30-RS		30	5	3	52	11
p20/s6-RS		6	Full-pene		18	6
p30/s30-RS	30	30	6	6	52	11
p30/s15-RS		15	6	6	33	9
p30/s6		6	Full-pene		18	6
p25/s14-RS	25	14	5	5	30	8
p25/s12-RS		12	5	5	28	8
p15/s14-RS	15	14	5	2	30	8
p15/p7-RS		7	5	2	21	7
p10/s30-RS	10	30	5	3	52	11
p10/s20		20	5	3	40	10
p10/s16		16	5	3	34	9
p6/s6-RS	6	6	Full-pene		18	6

TABLE 4.8.4

Stress Concentration Factors in Welded Joints with LTT Elongated Bead

	Stress Concentration Factor at 4 Positions			
Model-Name	**A**	**B**	**C**	**R**
p20/s16-RS	1.33	1.45	1.65	1.07
p20/s30-RS	1.39	1.07	1.35	0.98
p20/s6-RS	1.21	1.93	1.72	Full-pene
p30/s30-RS	1.41	1.23	1.51	1.11
p30/s15-RS	1.41	1.51	1.53	0.86
p30/s6	1.28	1.8	1.6	Full-pene
p25/s14-RS	1.41	1.47	1.48	0.89
p25/s12-RS	1.4	1.53	1.47	0.83
p15/s14-RS	1.41	1.27	1.32	0.87
p15/p7-RS	1.39	1.57	1.29	0.79
p10/s30-RS	1.36	0.76	0.98	0.63
p10/s20	1.4	1.38	1.3	0.83
p10/s16	1.39	1.55	1.41	0.9
p6/s6-RS	1.29	1.45	1.37	Full-pene

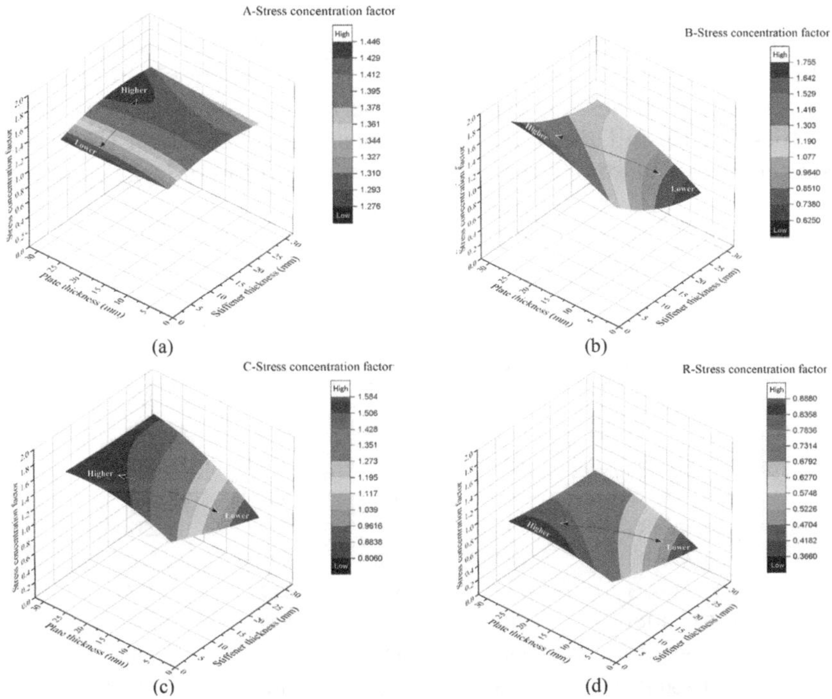

FIGURE 4.8.10
Stress concentration factor at weld toes A, B, C, and root R for various thicknesses t_p and t_s.

$$F = A_1 + A_2x + A_3y + A_4xy + A_5x^2 + A_6y^2 \quad (4.8.1)$$

where x and y are the main plate thickness t_p and stiffener thickness t_s. All stress concentration factors are less than 2.0. The factor at the weld to A increases with increasing thicknesses t_p and t_s. At the weld toes B and C, it increases with an increase in t_p and decreases with an increase in t_s. The stress concentration factor at the root R is small. Note that full-pene in the tables indicates the full penetration welding cases where the stiffener thickness is less than 7mm.

4.8.3.3 Residual Stress Evaluation

The welding residual stress σ_R at critical locations A, B, C, and R, calculated using the thermal–elastic–plastic finite element method for various numerical analysis models, is presented in Table 4.8.5. The analysis results show that large compressive stress is observed at locations A, B, and C in the LTT elongated bead section, while tensile residual stress is generated at the root R.

Based on the results from Tables 4.8.4 and 4.8.5, a comprehensive evaluation can be conducted on the risk of fatigue crack initiation at the weld toe (A, B), the intermediate section (C), and the root section (R) in LTT elongated bead joints. Assuming an average external load of 150MPa on the main plate, the total evaluation stress σ_T is defined as the sum of the concentrated stress σ_F, calculated from the stress concentration factor in Table 4.8.4, and the welding residual stress σ_R shown in Table 4.8.5. The results are shown in Table 4.8.6.

For all plate thickness combinations, the total evaluation stress σ_T at the weld toes (A, B) and the intermediate position (C) of the LTT elongated bead is compressive due to the significant contribution of large compressive

TABLE 4.8.5

Residual Stress in Welded Joints with LTT Elongated Bead

	Residual Stress σ_R [MPa] at 4 Positions			
Model-Name	A	B	C	R
p20/s16-RS	–639	–501	–894	226
p20/s30-RS	–631	–495	–675	210
p20/s6-RS	–623	–489	–947	Full-pene
p30/s30-RS	–769	–512	–826	312
p30/s15-RS	–743	–507	–901	231
p25/s14-RS	–763	–535	–901	200
p25/s12-RS	–765	–518	–919	217
p15/s14-RS	–663	–520	–800	256
p15/p7-RS	–681	–433	–790	199
p10/s30-RS	–516	–535	–636	235
p6/s6-RS	–529	–333	–591	Full-pene

TABLE 4.8.6

Summed Total Stress σ_T from External Stress σ_F and Residual Stress σ_R in Welded Joints with LTT Elongated Bead

	Total Evaluation Stress σ_T [MPa] at 4 Positions			
Model Name	A	B	C	R
p20/s16-RS	−440	−284	−647	387
p20/s30-RS	−423	−335	−473	357
p20/s6-RS	−442	−200	−689	Full-pene
p30/s30-RS	−558	−328	−600	479
p30/s15-RS	−532	−281	−672	360
p25/s14-RS	−552	−315	−679	334
p25/s12-RS	−555	−289	−699	342
p15/s14-RS	−452	−330	−602	387
p15/p7-RS	−473	−198	−597	318
p10/s30-RS	−312	−421	−489	330
p6/s6-RS	−336	−116	−386	Full-pene

residual stress, suggesting a low likelihood of fatigue crack initiation. On the other hand, the total evaluation stress σ_T at the weld root (R) is highly tensile, indicating a relatively higher risk of crack initiation.

4.8.3.4 Appropriate Penetration Parameters Determined from Stress Intensity Factor Evaluation in Different Thickness Combinations

Using the characteristic tensor method [36], the stress intensity factor K at the weld root R was calculated under combined residual stress σ_R and applied stress (150 MPa) induced concentration for the reference model corresponding to experimental specimens (t_p=20 mm, t_s=16 mm, a = 5 mm, b = 3 mm) and various combinations of thicknesses ranging from 6 to 30 mm with designed penetration depths a and b.

Using this K value, the required penetration parameter depths (a, b) to ensure the root stress intensity factor remains below the reference value were investigated for various plate thickness combinations. The results are presented in Figure 4.8.11 [32], where the numbers inside each frame represent the calculated stress intensity factor values, all of which are lower than that of the reference model. Note that full penetration in the figure indicates cases where the stiffener plate is thin ($t_s \leq 7$ mm), eliminating the root region entirely.

When both the main plate and stiffener are thick (30 mm each, t_p=30 mm, t_s=30 mm), a weld penetration depth of a=6 mm and b=6 mm yields a root stress intensity factor of 481 N/mm$^{3/2}$, which is lower than the reference value of 487 N/mm$^{3/2}$. This demonstrates that if LTT elongation bead welding is performed under the optimized reinforcement conditions (a, b) derived from numerical analysis, fatigue failure originating from the root can be suppressed.

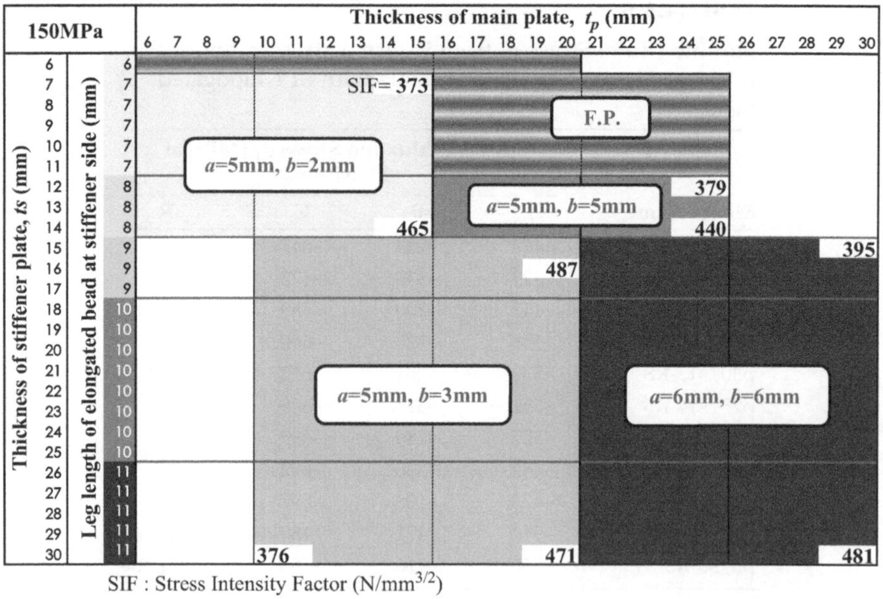

FIGURE 4.8.11
Penetration depths *a* and *b* for strengthening the root of the boxing fillet-welded joint with LTT elongated bead in different thickness combinations.

References

1. O. Watanabe, S. Matsumoto, Y. Nakano, and Y. Saito, Fatigue strength of welded joint of high strength steel and its controlling factor, effects of stress concentration factor and welding residual stress. *Quaternary Journal of the Japan Welding Society*, 13(3) (1995), 438–443 (in Japanese).
2. P.J. Haagensen and S.J. Maddox, *Post weld Fatigue Life Improvement of Steel and Aluminum Structures*, IIW document XIII-2200r7-07 (2010).
3. T. Hanji and K. Anami, Influence of static strength of steels on fatigue strength improvement of welded joints by toe treatment. *Journal of Structural Engineering*, 63A (2017), 646–656 (in Japanese).
4. B. Mecséri and B. Kövesdi, Assessment of grinding weld treatment methods using effective notch stresses. *Welding in the World*, 64 (2020), 1033–1046.
5. Y. Kotani, T. Tsuyama, A. Buerlihan, and S. Tsutsumi, Improvement of fatigue strength of out-of-plane gusset welded joints by controlled-additional weld bead shape and toe grinding. *Journal of Structural Engineering*, 67A (2021), 497–508 (in Japanese).
6. H. Al-Karawi, R.U.F.V.B.U. Polach, and M. Al-Emrani, Fatigue crack repair in welded structures via tungsten inert gas remelting and high frequency mechanical impact. *Journal of Constructional Steel Research*, 172 (2020), 106200.
7. S.J. Maddox, M.J. Dore, and S.D. Smith, A case study of the use of ultrasonic peening for upgrading a welded steel structure. *Welding in the World*, 55(9/10) (2011), 56–67.

8. T. Mori, H. Shimanuki, and M. Tanaka, Effect of unit on fatigue strength of web-gusset welded joints considering service condition of steel structures. *Welding in the World*, 56 (2012), 141–149.
9. A. Hobbacher, Recommendations for Fatigue Design of Welded Joints and Components, Doc. XIII-2151-07/XV-1254-07, International Institute of Welding, Paris, France, May (2007).
10. K. Ruiz, N. Osawa, and S. Rashed, A practical analysis of residual stresses induced by high-frequency mechanical impact post-weld treatment. *Weld in the World*, 63(5) (2019), 1255–1263.
11. H. Murata, N. Kato, and H. Tamura, Effect of phase transformation on welding residual stress, stress relaxation by transformation induced plasticity (Report 5). *Quarterly Journal of the Japan Welding Society*, 11(4) (1993), 545–550 (in Japanese).
12. A. Ohta, O. Watanabe, K. Matsuoka, C. Siga, S. Nishijima, Y. Maeda, N. Suzuki, and T. Kubo, Fatigue strength improvement by using newly developed low transformation temperature welding material. *Welding in the World*, 43 (1999), 38–42.
13. C. Miki, K. Anami, and Y. Higuchi, Fatigue strength improvement by additional welding with low temperature transformation welding material. *Journal of Structural and Construction Engineering*, 710(1–60) (2002), 311–319 (in Japanese).
14. M. Tai, C. Miki, and K. Suzuki, Fatigue strength improvement of out-of-plane gusset welded joints of steel girder made of high yield steel. *Journal of Japan Society of Civil Engineers, Ser. A1*, 68(2) (2012), 356–366 (in Japanese).
15. C. Shiga, H. Murakawa, K. Hiraoka, N. Osawa, H. Yajima, T. Tanino, S. Tsutsumi, T. Fukui, H. Sawato, K. Kamita, T. Matsuzaki, T. Sugimura, T. Asoda, and K. Hirata, Elongated bead weld method for improvement of fatigue properties in welded joints of ship hull structures using low transformation temperature welding materials. *Welding in the World*, 61 (2017), 769–788.
16. A. Hobbacher, *Recommendations for Fatigue Design of Welded Joints and Components*, IIW document XIII-2151r1-07/XV-1254r1-07 (2007).
17. C. Shiga, H. Murakawa, Y. Matsuo, U. Ohsuga, K. Hiraoka, Y. Morikage, and K. Yasuda, Fatigue improvement in high-strength steel welded joints with compressive residual stress. *Welding in the World*, 58 (2014), 55–64.
18. C. Shiga, H.Y. Yasuda, K. Hiraoka, and H. Suzuki, Effect of M_s temperature on residual stress in welded joints of high-strength steels. *Welding in the World*, 54(3/4) (2010), 71–79.
19. C. Shiga, K. Hiraoka, H. Murakawa, N. Osawa, H. Yajima, and T. Tanino, Improvement of fatigue life by elongated bead method with low transformation temperature material. *Welding Technology*, 1 (2016), 66–72 (in Japanese).
20. C. Shiga, K. Hiraoka, H. Yajima, T. Tanino, H. Murakawa, S. Tsutsumi, and N. Osawa, Development of advanced fatigue strength improvement method for ship's out-of-plane gusset welded joints using low transformation temperature welding material. In *Proceedings of the Twenty-fifth (2015) International Ocean and Polar Engineering Conference Kona*, Big Island, Hawaii, USA, June 21–26 (2015).
21. T. Matsuzaki, K. Hiraoka, Z. Feng, N. Ma, H. Murakawa, S. Kano, K. Okada, S. Kimura, C. Shiga, and H. Yajima, New LTT welding material and all-position repair welding with elongated bead for super-long fatigue life of boxing fillet joints. *Journal of The Japan Society of Naval Architects and Ocean Engineers*, 32(365) (2020), 153–161 (in Japanese).
22. K. Saito, T. Hirashima, N. Ma, and H. Murakawa, Characteristic tensor for evaluation of singular stress field under mixed-mode loadings. *Computer Modeling in Engineering & Sciences*, 122(2) (2020), 415–432.

23. T. Matsuzaki, N. Ma, K. Hiraoka, H. Murakawa, N. Osawa, T. Takeuchi, K. Okada, C. Shiga, H. Yajima, and S. Kano, Verification of the fatigue life extension and practicality of the LTT elongated bead welding method for boxing fillet joints. *Journal of the Japan Society of Naval Architects and Ocean Engineers,* 40 (2024), 133–143 (in Japanese).
24. N. Ma, Z. Feng, K. Hiraoka, and T. Matsuzaki, Compressive residual stresses in LTT elongated bead welded in all positions for fatigue crack prevention of boxing fillet joints. *Journal of Manufacturing Processes,* 117 (2024), 82–94.
25. Z. Feng, N. Ma, K. Hiraoka, Y. Komizo, S. Kano, and M. Nagami, Development of 16%Cr–8%Ni low transformation temperature welding material for optimal characteristics under various dilutions due to all repair welding positions. *Science and Technology of Welding and Joining,* 28(4) (2023), 305–313.
26. Z. Feng, N. Ma, S. Tsutsumi, and F. Lu, Investigation of the residual stress in a multi-pass T-welded joint using low transformation temperature welding wire. *Materials,* 14(2) (2021), 325.
27. H. Huang, N. Ma, S. Rashed, K. Narasaki, K. Kubo, Y. Agano, N. Tanaka, and W. Mao, Experimental and numerical investigation of a gouging heat source model and gouging/welding residual stress and deformation. *Journal of Manufacturing Processes,* 110 (2024), 101–113.
28. M.B. Prime, Cross-sectional mapping of residual stresses by measuring the surface contour after a cut. *Journal of Engineering Materials and Technology,* 123 (2) (2001), 162–168
29. N. Ma, Z. Feng, K. Hiraoka, and T. Matsuzaki, Compressive residual stresses in LTT elongated bead welded in all positions for fatigue crack prevention of boxing fillet joints. *Journal of Mathematical Physics,* 117(15) (2024), 82–94
30. Z. Barsoum and M. Gustafsson, Fatigue of high strength steel joints welded with low temperature transformation consumables. *Engineering Failure Analysis,* 16(7) (2009), 2186–2194.
31. Z. Barsoum and M. Gustafson, Spectrum fatigue of high strength steel joints welded with low temperature transformation consumables. In *2nd International Conference on Fatigue Design,* Senlis, France. November 21–22 (2007).
32. N. Ma, H. Murakawa, K. Hiraoka, Z. Feng, W. Huang, S. Rashed, and T. Matsuzaki, FE analysis of LTT elongated bead method for boxing welded joints with various thicknesses. *Journal of The Japan Society of Naval Architects and Ocean Engineers,* 41 (2025), 45–52 (in Japanese).
33. N. Ma, An accelerated explicit method and GPU parallel computing for thermal stress and welding deformation for large structural models. *The International Journal of Advanced Manufacturing Technology,* 87(3–5) (2016), 2195–2211.
34. N. Ma and K. Narasaki, Simulation of welding thermal conduction and thermal stress using hybrid method of accelerated explicit and implicit FEM. *IOP Conf. Series: Journal of Physics: Conf. Series,* 1063(012073) (2018), 1–6.
35. Z. Feng, N. Ma, W.N. Li, K. Narasaki, and F.G. Lu, Efficient analysis of welding thermal conduction using the Newton–Raphson method, implicit method and their combination. *International Journal of Advanced Manufacturing Technology,* 111 (2020), 1929–1940.
36. K. Saito, T. Hirashima, N. Ma, and H. Murakawa, Crack singular field evaluation with characteristic tensor considering residual stress. *Quarterly Journal of the Japan Welding Society,* 39(2) (2021), 105–114 (in Japanese).

5

LTT Welding Materials Used in Wire-Arc Additive Manufacturing and Surface Cladding

5.1 LTT Welding Material Used in a WAAM Wall for Compressive Residual Stress

5.1.1 Descriptions of a WAAM Wall and LTT Welding Material

Figure 5.1.1 shows the experimental setup for a WAAM wall and its dimensions using LTT welding material, which consists of 13 layers [1]. The fixtures were employed during the WAAM process and were released when the temperature on the final layer surface dropped below 150°C.The WAAM conditions for each layer are 149 A for welding current, 14.5 V for welding voltage, 250 mm/min for torch moving speed, and 6.5 m/min for welding wire feeding speed, respectively. The torch moving path is of the weaving type with a frequency of 2 Hz and a weaving amplitude of 8 mm. The welding torch moved up 2.3 mm in the z-direction for the next layer deposition. During the depositing process, the interpass temperature was kept below 180°C.

The 10%Cr–10%Ni LTT welding wire whose Ms is 180°C was employed for deposition, and the base plate material is stainless steel SUS304, whose chemical compositions and Ms temperature are shown in Table 5.1.1. The information for the stainless welding wire SUS308L listed in the table is for comparison with the LTT material.

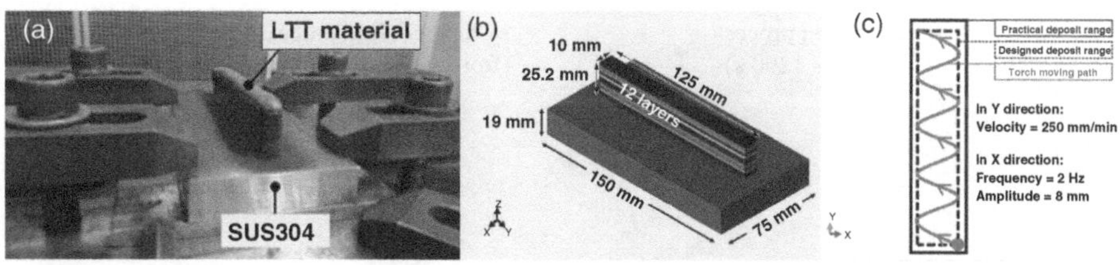

FIGURE 5.1.1
Experimental setup for a WAAM wall.

DOI: 10.1201/9781003601845-5

5.1.2 Finite Element Modeling for the WAAM Wall and Residual Stress

To investigate the possibility of compressive residual stress generation from the LTT welding wire in WAAMed components, finite element analysis needs to be conducted beforehand. Figure 5.1.2 shows the finite element mesh division. The minimum element size is 0.25 mm in the deposition height direction. Figure 5.1.3 shows the modeling of boundary condition changes during the WAAM process and cooling process. In the final cooling process, after the fixtures were released, only rigid motion of the WAAM wall model is constrained in the analysis.

TABLE 5.1.1

Chemical Compositions of SUS308L and LTT Welding Wire

	Chemical Compositions [wt%]						
Materials	**C**	**Si**	**Mn**	**Ni**	**Cr**	**Mo**	**Ms [°C]**
10%Cr–10%Ni LTT	0.026	0.320	0.07	10.0	10.0	0.13	180
SUS308L	0.024	0.85	1.58	10.6	19.91	–	–83

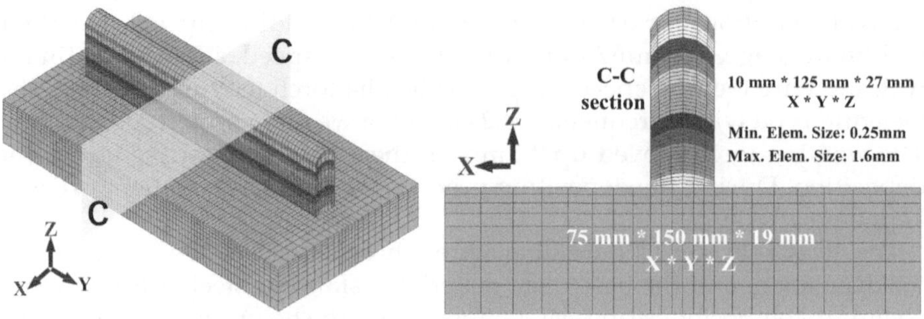

FIGURE 5.1.2
Finite element modeling for the WAAM wall.

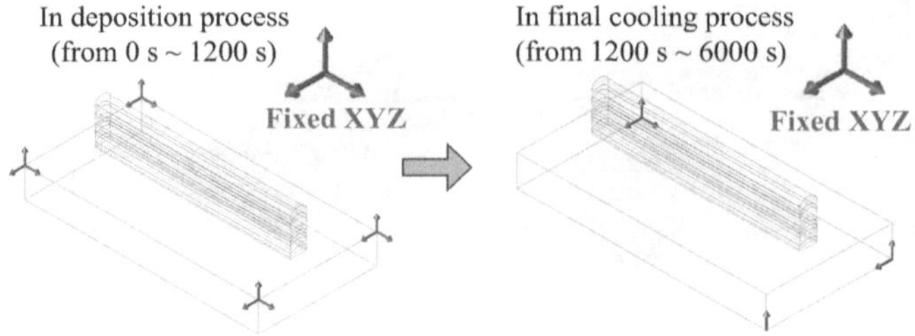

FIGURE 5.1.3
Modeling for boundary condition change during WAAM process and cooling process.

The analyzed longitudinal residual stress distribution in the WAAMed wall model and its comparison using 10%Cr–10%Ni LTT material and stainless SUS308L are shown in Figure 5.1.4. Large compressive residual stress can be observed in the top deposited layers of the LTT material. The detailed distribution on the middle transverse section and in the deposition height direction is shown in Figure 5.1.5. It is clear that compressive residual stress exists in the top four LTT deposited layers, with a height of about 7 mm. In contrast, with conventional stainless SUS308L, tensile residual stress exists in most deposited layers. To understand the mechanism of compressive residual stress generation in the WAAM process, the residual stress distribution and temperature distribution after each layer of deposition are illustrated in Figure 5.1.6. After depositing four layers, the compressive residual stress distribution reaches the quasi-steady state and increases with the deposition layers. The quasi-steady height of compressive residual stress is about 7 mm, corresponding to the height of three layers, which corresponds to the heat-affected zone (HAZ) of the LTT material. Specifically, in the HAZ, the reheating induced peak temperature due to the latest deposition layer is higher than the austenitization temperature (about 600°C).

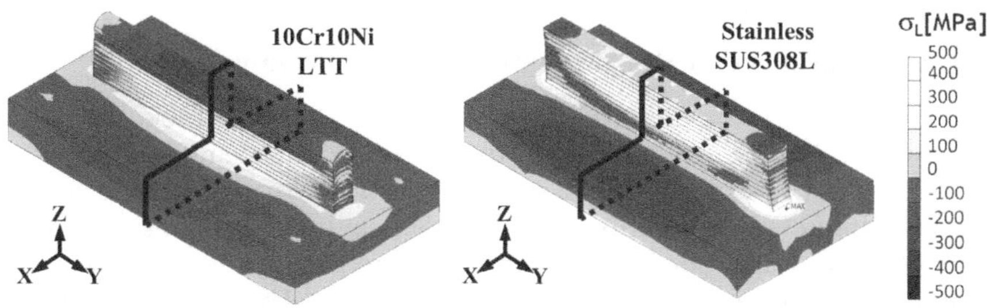

FIGURE 5.1.4
Residual stress comparison between LTT material and stainless SUS308L.

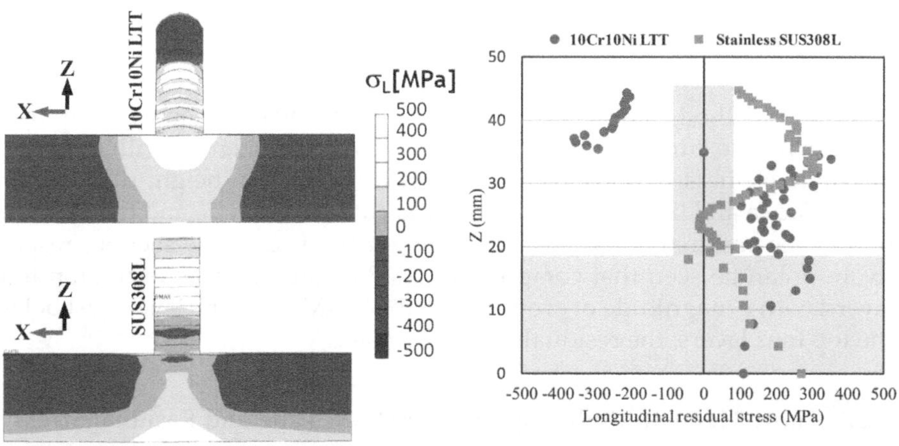

FIGURE 5.1.5
Internal residual stress comparison between LTT material and stainless SUS308L.

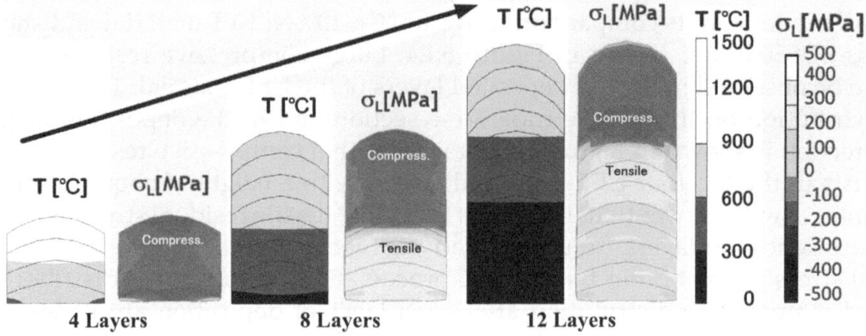

FIGURE 5.1.6
Relationship between the compressive residual stress zone and the HAZ in the quasi-steady state.

5.2 LTT Welding Material Used in a WAAM Tube for Compressive Residual Stress

5.2.1 Descriptions of a WAAM Tube Using LTT Welding Material

Figure 5.2.1 shows the experimental setup for a WAAM tube and related dimensions [2]. The materials of the WAAM tube and substrate are 10%Cr–10%Ni LTT material and stainless SUS304, respectively. The inner and outer diameters of the tube are 74 and 83 mm (wall thickness 6 mm). The height of the WAAM tube is 27 mm and was deposited in 20 layers. The average thickness for each deposited layer is 1.35 mm. The WAAM conditions are 149 A for welding current, 14.5 V for welding voltage, 6.5 m/min for wire feeding speed, and 16.7 mm/s for torch moving speed.

5.2.2 Measurement of Residual Stress and Its Distribution in the WAAM Tube

A contour method [3] involving cutting the transverse section was employed to measure the circumferential residual stress component σ_c with the aid of the finite element method using the measured section height distribution. Figure 5.2.2a and b shows the photo of the cut WAAM tube and the distribution of the residual stress component σ_c on the transverse section, respectively. It can be seen that compressive residual stress exists in the top four layers, with a magnitude of around –200 to –300 MPa. In the layers just below the top four layers, the residual stress is tensile.

5.2.3 Finite Element Modeling for the WAAM Tube and Residual Stress

To investigate the residual stress generation mechanism in the WAAM tube, finite element model for thermal–elastic–plastic analysis, as shown in

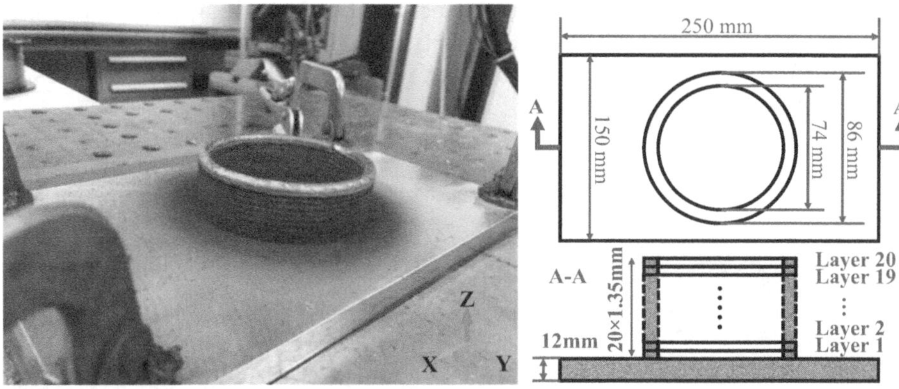

FIGURE 5.2.1
Experimental setup for a WAAM tube and related dimensions.

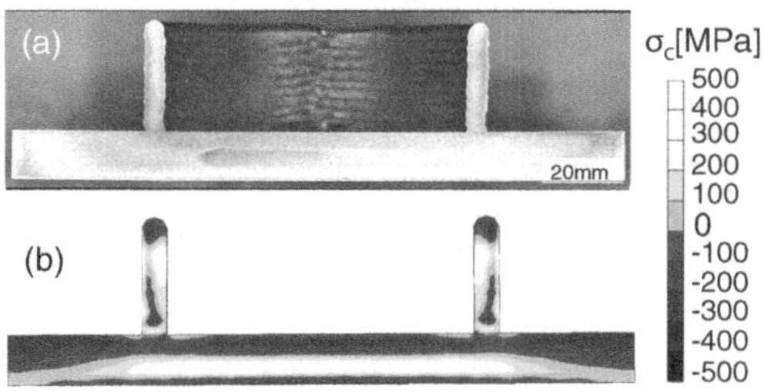

FIGURE 5.2.2
The cut WAAM tube and measured distribution of residual stress component σ_c.

Figure 5.2.3, was developed. The minimum mesh size is 0.3 mm in the deposition layers.

The analyzed residual stress distribution along the height direction and its comparison with the measured one are shown in Figure 5.2.4. Both the analysis and measurement indicate that compressive residual stress was generated in the top four layers of the WAAM tube.

Figure 5.2.5 shows the changes in residual stress with the increase of deposition layers. When the first four layers were deposited, compressive residual stress was generated in all four layers. However, with continuous deposition from the 5th layer to the 20th layer, the compressive residual stress exists only in the latest four layers, which have a depth of about 6 mm, corresponding to the HAZ of the LTT material. The residual stress distribution reaches a quasi-steady state after the fifth layer as the deposition layers increase. This occurs because the heat from the current deposition layer reheated the previous three layers to above the austenitization temperature (about 600°C), as shown in Figure 5.2.6, resulting in martensitic transformation occurred together within these four layers. Based on this knowledge, to make compressive residual stress in WAAM parts, the LTT material can be actively used for the deposition of the first four, for the most recent latest layers, or selectively used for some layers.

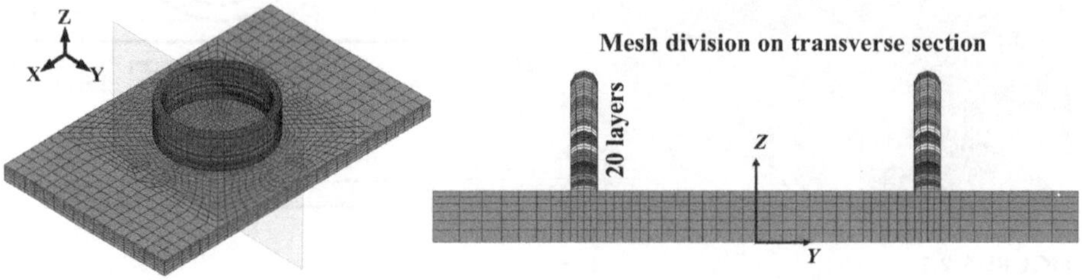

FIGURE 5.2.3
Finite element modeling for a WAAM tube in residual stress analysis.

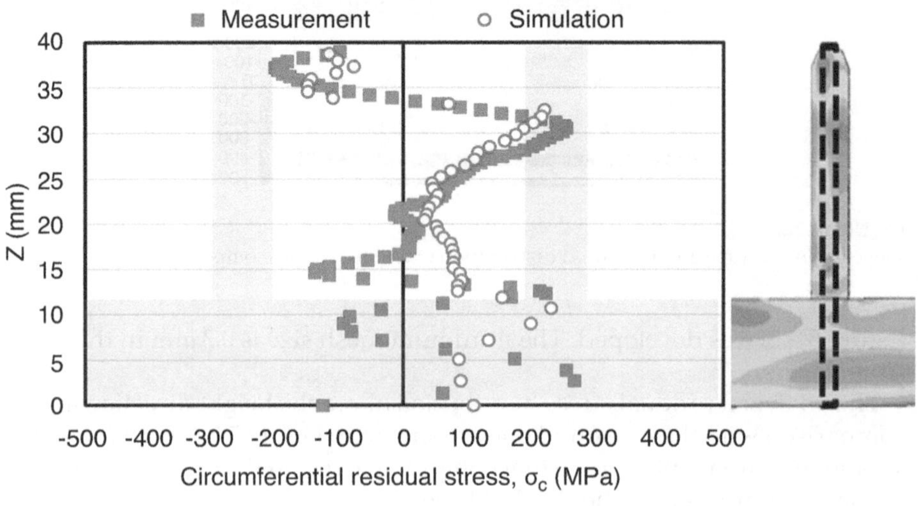

FIGURE 5.2.4
Comparison of measured and analyzed residual stress distribution in the height direction.

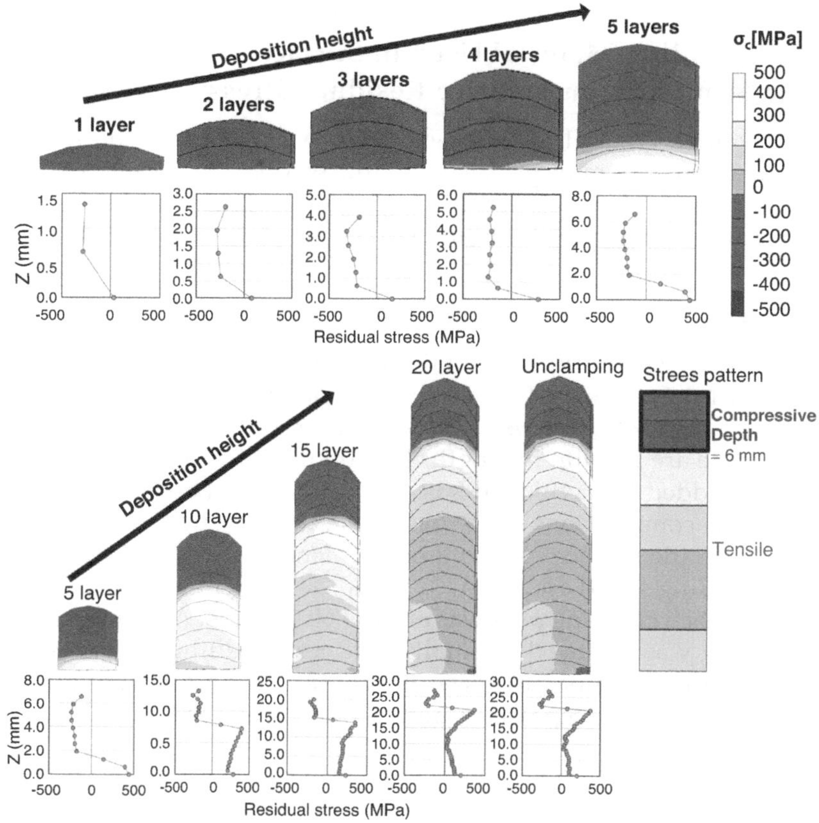

FIGURE 5.2.5
Analyzed residual stress and its change in a WAAM tube with different height.

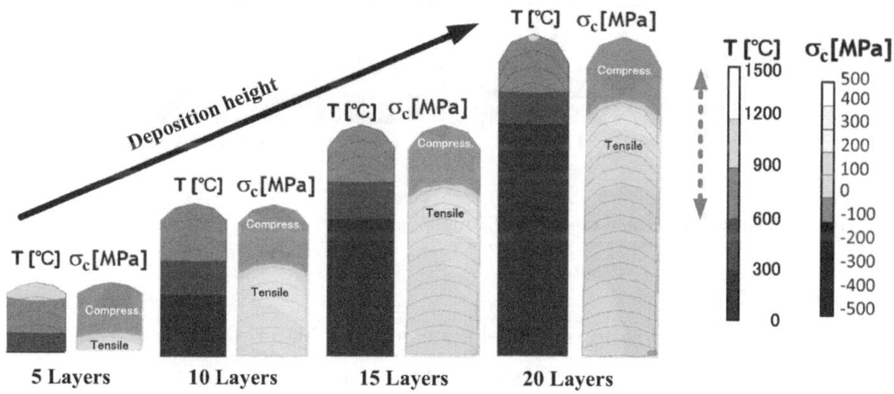

FIGURE 5.2.6
Relationship between compressive residual stress zone and HAZ in quasi-steady state.

5.3 LTT Welding Material Used in Surface Cladding for Compressive Residual Stress

In the previous sections, LTT welding material was used for deposition in the height direction with multiple layers. In this section, the potential for generating compressive residual stress in wide-area surface cladding is investigated through both experiments and numerical analyses [4].

5.3.1 Surface Cladding Experiment and LTT Material

Figure 5.3.1 shows the experimental setup for the surface cladding of LTT material. A 16%Cr–8%Ni LTT material was cladded onto the surface of the steel plate KA36, which is used in welded structures. The dimensions of the steel plate are 200 mm in length, 100 mm in width, and 20 mm in thickness. The cladded layer thickness is 4 mm. Totally, 8 passes were cladded. The chemical compositions of the base plate and LTT material are listed in Table 5.3.1. In the surface cladding experiment, the welding current, voltage, and welding speed are 200 A, 32 V, and 200 mm/min, respectively. The arc time for each cladding pass was around 60 seconds. After each cladding pass, the slag was removed manually and next cladding pass was continued. The interpass temperature was not controlled, and the measured data during the cladding of 8 passes is between 173°C and 360°C (Table 5.3.2).

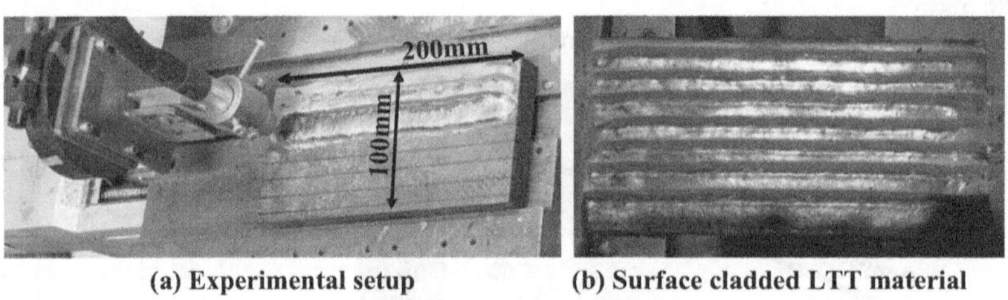

FIGURE 5.3.1
Experimental setup for surface-cladded LTT material.

TABLE 5.3.1
Chemical Compositions of Steel Plate and LTT Material

Materials	C	Si	Mn	P	S	Ni	Cr
KA36 plate	0.15	0.28	1.10	0.013	0.003	–	–
16%Cr–8%Ni LTT	0.02	0.34	0.44	0.02	0.01	8.2	16.1

TABLE 5.3.2

WAAM Conditions for Surface Cladding

Pass No.	Current [A]	Voltage [V]	Speed [mm/min]	Cladding Time [sec]	Interpass Temperature [°C]
1	200	32	200	59	36
2	200	32	200	59	173
3	200	32	200	60	303
4	200	32	200	60	360
5	200	32	200	58	281
6	200	32	200	59	301
7	200	32	200	60	299
8	200	32	200	61	300

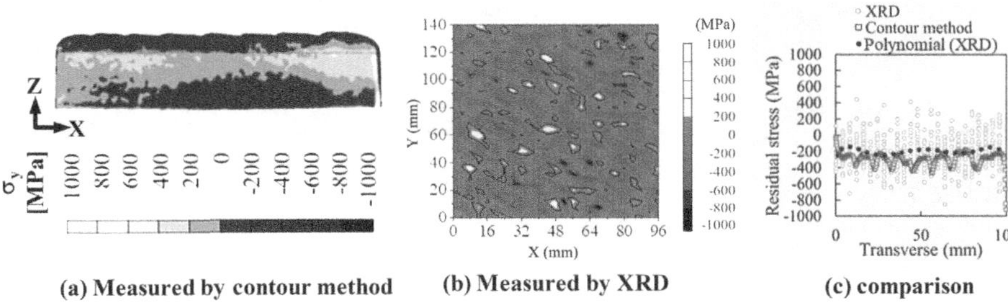

FIGURE 5.3.2
Measured residual stress on the transverse section and on the cladded surface.

5.3.2 Measurement of Residual Stresses in Surface Cladded Layer

The residual stress distribution along the transverse section measured by the contour method [3] is shown in Figure 5.3.2a. It can be seen that large compressive residual stress was generated in the cladded layer. Below the cladded layer, tensile residual stress exists. In addition to the measurement using the contour method, the residual stress on the cladded surface was measured by the XRD method and is shown in Figure 5.3.2b. Although there are significant variations, compressive residual stress can be clearly observed. Figure 5.3.2c shows the surface residual stress comparison between the two measurement methods. The averaged residual stress measured by XRD and the contour method shows good consistency.

5.3.3 Finite Element Modeling for LTT Surface Cladding and Residual Stresses

To investigate the three-dimensional residual stresses induced by surface cladding, a finite element model was developed as shown in Figure 5.3.3, and a thermal–elastic–plastic analysis was conducted, as shown in Figure 5.3.4. The analyzed results show that both the longitudinal residual stress σ_y and the transverse one σ_x are compressive in the cladded layer. A similar distribution through the thickness direction can also be seen in Figure 5.3.5. These isotropic compressive residual stresses may enhance the practical applications.

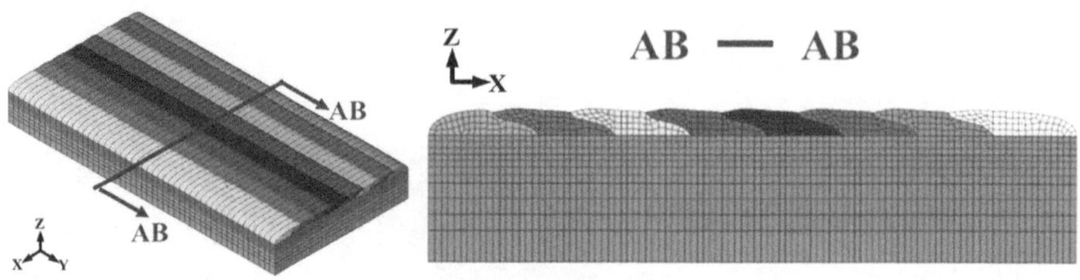

FIGURE 5.3.3
Finite element model.

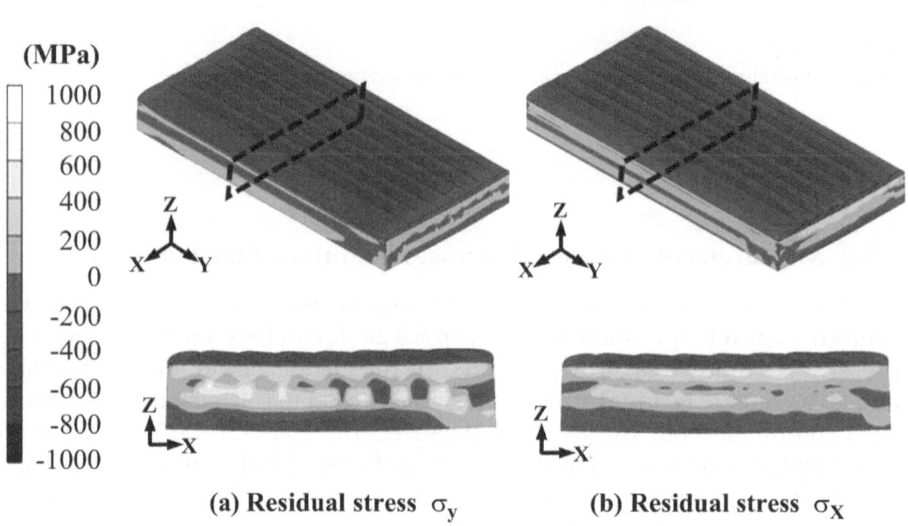

FIGURE 5.3.4
Comparison of analyzed residual stresses σ_y and σ_x in surface cladded test pieces.

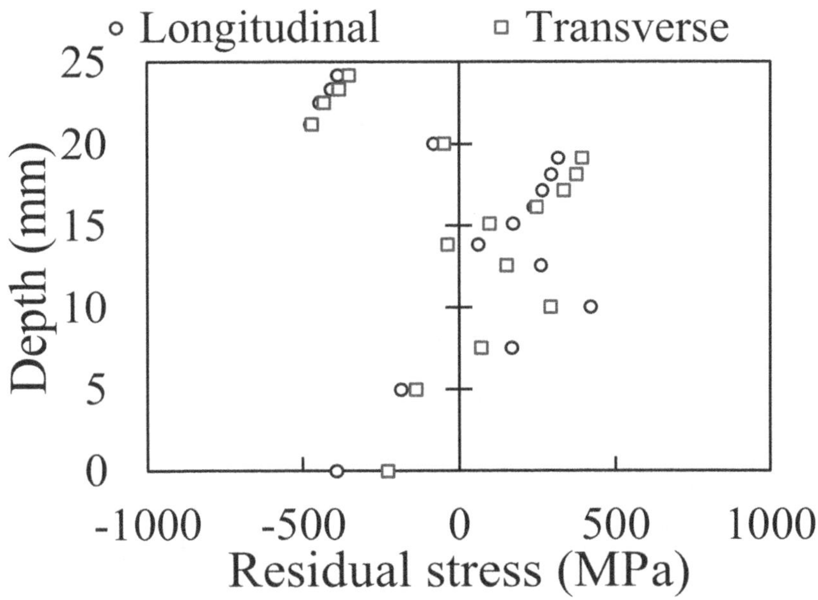

FIGURE 5.3.5
Comparison of residual stresses σ_y and σ_x distributed through the thickness direction.

5.4 LTT Welding Material Used in WAAM Components for Distortion Reduction

In previous chapters, the compressive residual stress induced by LTT welding materials was demonstrated. In this chapter, the experimental results of LTT welding wire for reducing the distortion of WAAM components are introduced [5].

5.4.1 Materials and WAAM Conditions

LTT welding wire (10%Cr–10%Ni) and stainless welding wire (SUS308L) were employed for a WAAM wall model, where the substrate is austenitic stainless steel SUS304. The chemical compositions and the Ms temperature of the 10%Cr–10%Ni LTT welding wire and SUS308L stainless welding wire are shown in Table 5.1.1. The target geometry is shown in Figure 5.4.1. The target dimensions in the *x*, *y*, and *z* directions are approximately 125, 10, and 25 mm, respectively.

The experimental setup and fixtures are shown in Figure 5.4.2. The fixtures were released when the temperature on the final layer surface was lower

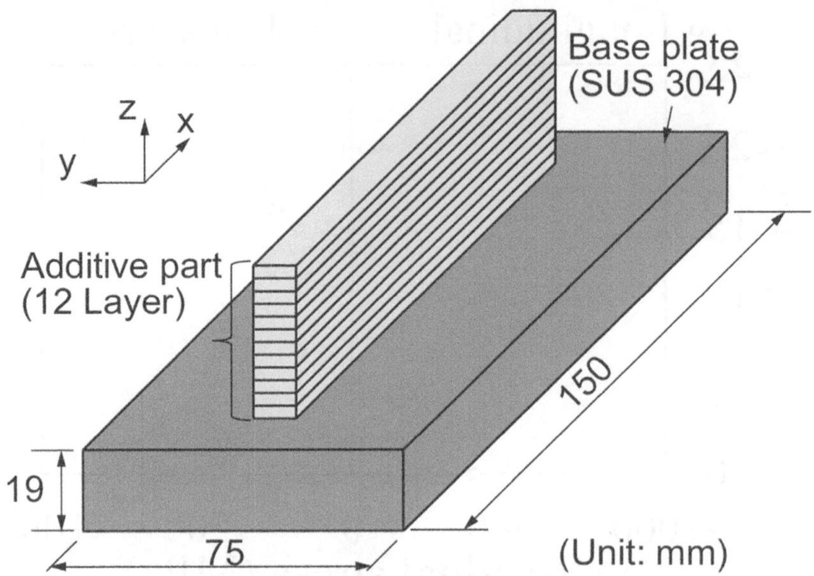

FIGURE 5.4.1
Target geometry of a wall model.

FIGURE 5.4.2
Experimental setup for WAAM.

than 150°C. The wire arc discharge current (welding current or deposition current), voltage, torch moving speed, and metal wire feed speed were 149 A, 14.5 V, 250 mm/min, and 6.5 m/min, respectively. To maintain a constant thickness and width for each layer, an arc weaving operation was performed. The weaving amplitude was 8 mm, and the frequency was 2 Hz. The welding torch moved up 2.3 mm in the z direction for next layer deposition, regardless of the welding wires used. While depositing SUS308L layers, the interpass temperature was controlled to be less than 150°C. For depositing LTT layers, the interpass temperature was kept below 180°C.

5.4.2 Experimental Cases

In the experiments, a total of six test pieces shown in Figure 5.4.3 were WAAMed using SUS308L welding wire and 10%Cr–8%Ni LTT welding wire in different layers. Case 1 used the SUS308L welding wire for all 12 layers. In Case 2, SUS308L wire was used for the first 8 layers, and LTT wire was deposited for the final 4 layers. In Case 3, SUS308L was deposited in layers 1–4 and 9–12, while the LTT wire was used in middle layers 5–8. Case 4 shows that the first 4 layers were deposited using LTT wire and the final 8 layers used SUS308L wire. Case 5 is similar to Case 4, but the first 6 layers, and the final 6 layers used LTT and SUS308L wires. Finally, in Case 6, all 12 layers were deposited using LTT wire.

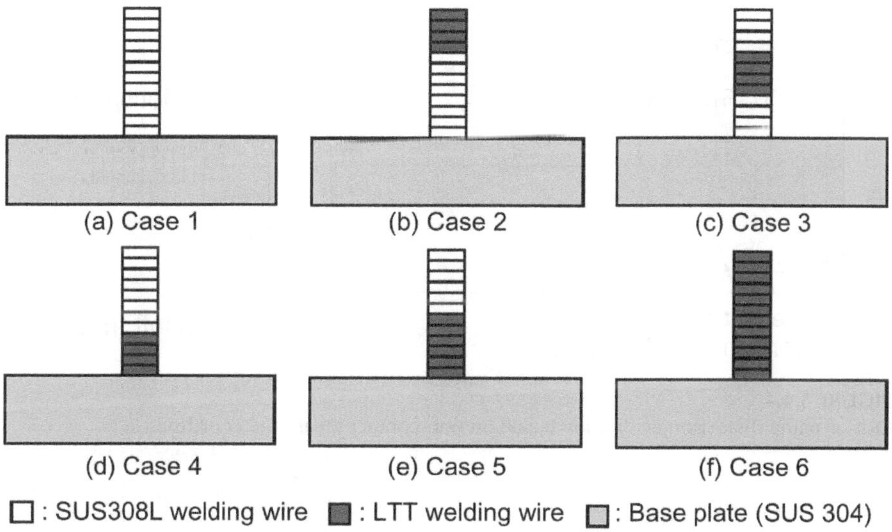

FIGURE 5.4.3
Experimental cases for WAAMed walls.

5.4.3 WAAM Distortion Comparison

The distortion of the base plates of the six test pieces was evaluated by a 3D non-contact coordinate measurement system shown in Figure 5.4.4. The distortion contour for all 6 cases is shown in Figure 5.4.5. For quantitative comparison, the curvature ρ for longitudinal bending distortion of the base plates of the test pieces along the x-axis is derived as shown in Figure 5.4.6. The deformation angle for transverse bending along the y-axis shown in Figure 5.4.7 was calculated from the measured data.

The longitudinal distortion when using LTT welding wire partially or fully is smaller than that of the test pieces made only with SUS308L wire. Especially, the longitudinal bending distortion of the base plate of the test piece Case 6, which was fully built up with LTT welding wire, is reduced drastically.

The angular distortion of the base plates exhibited limited change even when the test piece was fully built up with LTT welding wire. This is because the martensitic transformation-induced thermal expansion in the width direction of the first deposition layer has a larger contribution to the angular distortion of the base plates, while the contributions of other deposition layers mainly affect the local dimensional changes of the deposition layers. Another reason is that the bending stiffness along the z–x section (y=0) is much larger than that of the y–z section (x=0) with the increase in deposition layers.

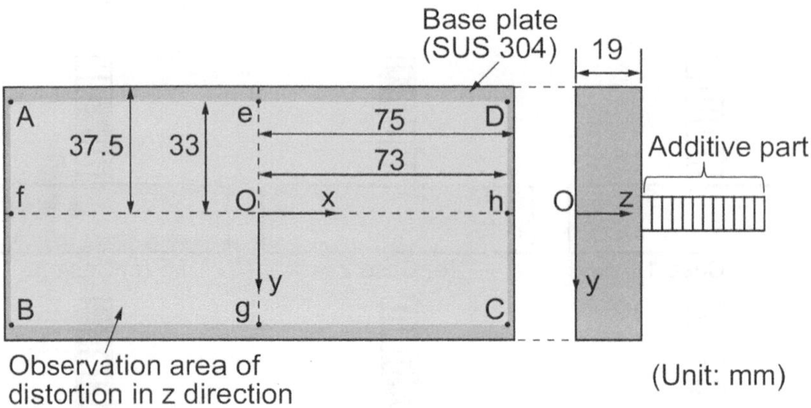

FIGURE 5.4.4
Out-of-plane distortion evaluation based on non-contact measured coordinates.

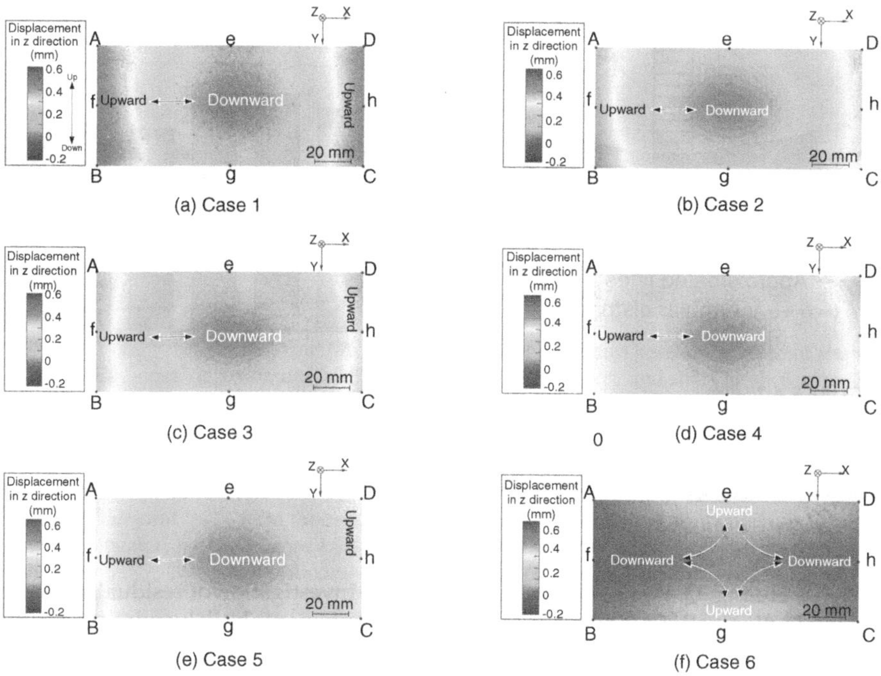

FIGURE 5.4.5
Measured out-of-plane displacement z and its distribution.

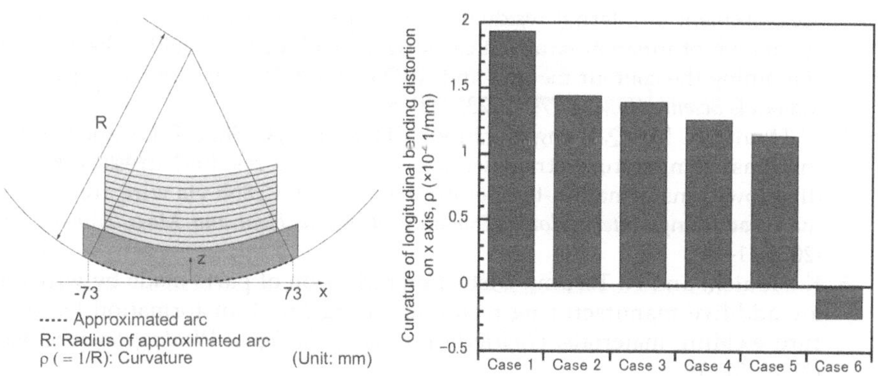

FIGURE 5.4.6
Quantitative evaluation method for longitudinal bending.

```
*PART
    Part definitions (links materials to sections)
$
$------------------------------------------------------------------------------------$
$       SET BLOCK
$------------------------------------------------------------------------------------$
$
$
*SET_NODE
    Defines applied forces or boundary conditions.
$
$------------------------------------------------------------------------------------$
$       NODE BLOCK
$------------------------------------------------------------------------------------$
$
$
*NODE
    Defines the coordinates of the nodes (grid points).
$
$------------------------------------------------------------------------------------$
$       ELEMENT XX BLOCK
$------------------------------------------------------------------------------------$
$
$
*ELEMENT_XX (e.g., *ELEMENT_SHELL, *ELEMENT_SOLID)
    Finite element connectivity (node IDs, part assignments)
$
*END
```

Index